P9-EMO-239

Praise for the First Edition of *Presenting to Win*

"I've taken the training. If you pay attention to what Jerry Weissman tells you (and it's hard not to), you'll be a measurably better public speaker. And that leads directly to success."

Stewart Alsop, (former) *Fortune* columnist, and Venture Partner, New Enterprise Associates

"Building a great company takes many years. Ruining its prospects can take less than an hour. *Presenting to Win* will let you tell the story right."

Alex Balkanski, General Partner, Benchmark Capital

"Read this book and get a master's degree in effective communications in one afternoon. The executive teams that we have invested with have used Weissman's communication techniques to dramatically improve team productivity and shareholder value."

David F. Bellet, Retired Chairman, Crown Advisors International, Ltd.

"Clear, concise, and high-impact communications are necessary for every business executive in today's high-speed world. Jerry Weissman provides leaders with simple tools they can draw on quickly in order to maximize the return on all of their communications efforts."

Sue Bostrom, Executive Vice President, Chief Marketing Officer, Global Policy and Government Affairs, Cisco Systems, Inc.

"Jerry Weissman is an expert in helping leading technology executives improve the clarity and substance of their communication. Jerry does an excellent job of giving sound recommendations that will lead to better communications and leadership."

James W. Breyer, General Partner, Accel Partners

"Jerry transformed my presentation skills by helping me think of the WIIFY and Aha! factor, and this helped me take my company public in 1998. Today, through this book, you can acquire the same skills on 'how to present to win' for a fraction of what it cost us to learn from Jerry. It is a must for professional success."

K. B. Chandrasekhar, Founder, Exodus Communications, and CEO, Jamcracker, Inc

"Jerry Weissman's approach is light years ahead of the rest of the world. My guess is that *Presenting to Win* will become a classic within its genre. Great job!"

Thomas M. Claflin II, Chairman, Claflin Capital Management, a Boston-based venture capital firm

"Inspiring leaders must possess great communication skills. They are one of the essential ingredients used to mobilize followers and move nations and markets. This book tells readers how to develop those skills."

Bill Davidow, Founding Partner, Mohr, Davidow Ventures, and author of *Marketing High Technology*, *Total Customer Service*, and *The Virtual Corporation*

"During one of the most important periods of my career, Jerry used the concepts in *Presenting to Win* to prepare me and my team for the EarthLink IPO road show. He helped us hone our message so that it could be crisply communicated to our audience, and in the process, we gained a better understanding of our own strategy than ever before. Jerry teaches you how to tell a story, which is more critical to success than most people think."

Sky Dayton, Founder, EarthLink and Boingo Wireless, and CEO of Helio

"Jerry's classic is the definitive encapsulation of a true 'maestro' playing his 'A' game. His techniques were an absolutely critical success factor in Luminous raising $80mm in one of the toughest market environments I've ever seen. There isn't a successful entrepreneur I know who won't benefit from leveraging his persuasive, maximum-impact presentation principles."

Dixon R. Doll, Founder and Managing General Partner, Doll Capital Management

"Young entrepreneurs with deep domain knowledge are frequently unable to organize and present their vision, or their company's story, in an orderly, crisp, and compelling fashion. Jerry Weissman has armed a succession of evolving executives with those critical skills. This seminal book cogently describes how he does it. *Presenting to Win* is an essential read for any CEO who intends to raise capital, wishes to elevate the visibility of his/her company in the financial community, or aspires to more credible communication with the company's employees."

Irwin Federman, General Partner, U.S. Venture Partners

"Practical and sound advice on how to communicate effectively. Critical for effective managers. This advice is powerful whether you are selling a project or a company."

Allan Ferguson, Retired Senior Partner, 3i Group PLC

"For many years, I have witnessed Jerry's magic as he helped our portfolio companies to formulate and communicate their unique selling propositions in a style that optimized the appeal to investors. It was not until I read his book that I started to see how valuable it would be to apply his methods personally in presentations that I make regularly to my partners and to our investors. I would think that everyone who does business presentations could benefit from this easy-to-read, easy-to-apply book."

Flip Gianos, General Partner, Interwest Partners

"It was a pleasure to read *Presenting to Win*. Jerry has been a master for so many of our CEOs from Alpha Partner days to current AVI. His book should provide excellent background and conditioning to any executive in need of clear and persuasive communication skills. It carries the reader from 'point A' to 'point B' in a coherent and thoughtful manner. I should hope that any reader will feel capable and compelled to adopt the methodology in their next presentation to win!"

Brian J. Grossi, General Partner, AVI Capital, L.P.

"The perfect presentation has the potential to impact the most important events in your company's life: financings, mergers, partnerships. Despite this, most executives and entrepreneurs continually fall short when it comes to delivering a crisp, clear message in a compelling and concrete fashion. Luckily for them, the master, Jerry Weissman, has decided to share his secrets in *Presenting to Win*."

Bill Gurley, General Partner, Benchmark Capital

"Compelling, no-nonsense script to clear, winning communication! A must-have."

Grant Heidrich, Partner Emeritus, Mayfield Fund

"Learn from the master on how to craft and deliver an effective, winning presentation. I have made over a thousand presentations over the past decade and have found Jerry's methodology and tips, which he outlines in his book, of tremendous value."

Jos Henkens, (former) Managing Director, Advanced Technology Ventures

"Jerry Weissman's presentation coaching is world class. A person contemplating a presentation without utilizing his expertise is like a soldier going to battle without weapons."

Jay C. Hoag, Founding General Partner, Technology Crossover Ventures

"Perception is often as, if not more, important than reality. Great presentation is more important than great substance in creating that perception. I don't know anybody who has more experience than Jerry in creating great perceptions."

Vinod Khosla, Affiliated Partner, Kleiner Perkins Caufield & Byers, and Founder, Khosla Ventures

"Jerry's help has been of enormous value to many of my companies over the years (at Venrock). Jerry has a unique ability to guide management teams to clearly identify and convey their key messages in a manner best suited to obtain the desired impact on their target audience. The transformation can be magical at times, and we have found that not only does the management team appreciate his advice, but the target audience also, as they are able to understand the key issues faster and more clearly than otherwise possible."

Patrick Latterell, Founder and General Partner, Latterell Venture Partners

"I read Jerry's book with great interest. The content is both interesting and informative, and the writing is easy to understand and read. Jerry is a pro's pro. If there were a MBA in Power Presentations, Jerry's book would be the bible!"

Mel Lavitt, (former) Vice Chairman, C.E. Unterberg Towbin

"An absolute necessity in your arsenal to gain an edge in the capital-raising battle."

Andrew J. Malik, Chairman, Needham & Company, LLC

"Jerry Weissman has had an enormous impact on the companies that I've managed: First Tuesday and now Ariadne Capital. Through his sessions and techniques, he's helped me to bring everyone onto the same page with a clear statement of what we're trying to achieve and to get credit for what we're doing by telling a better story. His latest book is testament to the fact that 'Form is Content.'"

Julie Meyer, CEO, Ariadne Capital, European venture capital group

"Jerry's book *Presenting to Win* is, like his consulting practice, totally on-point and practical. Jerry has worked with many of the companies I took public as an investment banker or invested in as a venture capitalist. He made a huge difference in each and every situation. The best evidence is that even the most polished CEOs in the valley made dramatically more effective presentations with his expert guidance. Since Jerry can't work with everyone individually, it is terrific to have his approach and thoughts available in *Presenting to Win*."

J. Sanford Miller, General Partner, Institutional Venture Partners, and Director, Vonage Holdings Corporation

"Jerry's methods in *Presenting to Win* helped us raise our own first venture fund with twice the commitments we needed in half the time. Those same methods then helped many of our portfolio companies not just in their presentations, but had a significant impact in the marketing of their products."

Andy Rachleff, Partner, Benchmark Capital

"*Presenting to Win* should be required reading for any business executive who will be presenting to any group of internal or external constituents. Having been a technology investment banker for nearly 15 years, I've seen thousands of road show and company presentations, and many of them failed to incorporate the tips and techniques outlined in *Presenting to Win*. I've sent dozens of clients to Jerry over the years, and they have always emerged with crisper, clearer, and more effective presentations."

Michael J. Richter, (former) Managing Director, Investment Banking CIBC World Markets

"Weissman's organization of a CEO's thoughts and presentation usually created 15-25% more value to the company from institutional investors."

Sanford Robertson, Founder and Partner, Francisco Partners

"Jerry Weissman always knew how to get the best out of a presentation. This book is the next best thing to hiring him as an advisor, and a lot less expensive."

Dick Spalding, venture capitalist, entrepreneur, and former client

"Jerry helps you focus on the substance of your presentation rather than just its form. He forces you to think clearly and succinctly. As a result, not only does your audience understand your business better, you do too."

Christopher Spray, Senior Partner, Atlas Venture

"Jerry's concepts on visual presentation techniques are simple and yet extremely powerful. They make the difference in getting the message across in a constrained time limit. A must-read for all high-tech managers."

Anthony Sun, Managing General Partner, Venrock Associates

"After you spend years building your startup, there's a fleeting moment in time where perception meets reality in the creation of the market value around it. There's not a person in the Valley that has coaxed more multiple points out of the market at that point than Jerry. His coaching on how to really efficiently organize and explain ANY subject matter is proven, and shows in his book."

Bill Tai, General Partner, Charles River Ventures

"Many people don't know who Jerry Weissman is or why he's known as 'The Wizard of Aaahs.' However, for those of us who've had a chance to work with him or be coached by him, he is revered as 'the professor of effective presentations.' *Presenting to Win* identifies the simple rules of a master and communicates those rules in a refreshing way. Jerry shares personal experiences with industry icons that help the reader see the practical side of the process. You'll quickly conclude that you'd really like to have Jerry on your team for your next big presentation!"

T. Peter Thomas, Cofounder and Managing Director, ATA Ventures

"Value simply cannot be realized unless it is communicated. Untold wealth has died on the vine because it was not effectively presented to key constituents in the value chain. In this respect, Mr. Weissman and his methods are responsible for the realization of tremendous value in the American economy over the past 15 years. Nobody teaches the art and science of business presentations better than Mr. Weissman. He is more than the best in his class . . . he created the class. I have recommended him to dozens of companies. With his help, they have dramatically improved the effectiveness of their internal and external presentations. In *Presenting to Win*, he makes his powerful techniques accessible to every person who cares about truly realizing value in their business presentations."

David M. Traversi, Managing Director, 2020 Growth Partners, LLC

"There are those rare people who can stand up in front of an audience and deliver an impromptu, high-impact message. For the rest of us, there is Jerry Weissman, who helps us create a high-impact presentation. His methodology lets you go beyond the jargon of any profession and develop a presentation that will be meaningful to your audience. Miraculously, your confidence will soar. He helped me and many others in business . . . and I am glad to see that he wrote a book. Beyond the logic of preparing presentations using his methods, the book is full of little nuggets that I will want to refer to over and over again. And so will you."

Les Vadasz, (former) Executive Vice President, Intel Corporation

"We pay big bucks to get Jerry to coach our company CEOs on how to tell their story, and it's worth every penny. Now everyone can learn about Jerry's secret sauce for the price of a few lattés."

Barry Weinman, General Partner, Associated Venture Investors, and Cofounder of Allegis Capital

"This is a practical guide to creating persuasive presentations, a skill required of most contemporary business professionals. Jerry makes it easy to learn how to craft the message and deliver the punch that gets the audience to agree with the premise: getting to Point B. More powerful than features to benefits, understanding the audience's WIIFY (what's in it for you) is a vital part of winning presentations. Should be read and used as a reference guide for those who seek success as a result of doing their homework and building a presentation that works. Jerry has trained a number of the AVI portfolio company CEOs and CFOs prior to their IPO road shows. His methodology works extremely well."

Peter L. Wolken, General Partner, AVI Management Partners, Special Limited Partner, Diamondhead Ventures

Presenting to Win

Presenting to Win

The Art of Telling Your Story

Updated and Expanded Edition

Jerry Weissman

Vice President, Publisher: Tim Moore
Associate Publisher and Director of Marketing: Amy Neidlinger
Acquisitions Editor: Jennifer Simon
Editorial Assistant: Heather Luciano
Editorial Consultant: Karl Weber
Art Consultant: Nichole Nears
Operations Manager: Gina Kanouse
Digital Marketing Manager: Julie Phifer
Publicity Manager: Laura Czaja
Assistant Marketing Manager: Megan Colvin
Marketing Assistant: Brandon Smith
Cover Designer: Chuti Prasertsith
Managing Editor: Kristy Hart
Senior Project Editor: Lori Lyons
Copy Editor: Gayle Johnson
Proofreader: Kay Hoskin
Senior Indexer: Cheryl Lenser
Senior Compositor: Gloria Schurick
Manufacturing Buyer: Dan Uhrig

Publishing as FT Press
Upper Saddle River, New Jersey 07458

FT Press offers excellent discounts on this book when ordered in quantity for bulk purchases or special sales. For more information, please contact U.S. Corporate and Government Sales, 1-800-382-3419, corpsales@pearsontechgroup.com. For sales outside the U.S., please contact International Sales at international@pearson.com.

Printed in the United States of America

First Printing November 2008

ISBN-10: 0-13-714417-2
ISBN-13: 978-0-13-714417-4

Pearson Education LTD.
Pearson Education Australia PTY, Limited.
Pearson Education Singapore, Pte. Ltd.
Pearson Education North Asia, Ltd.
Pearson Education Canada, Ltd.
Pearson Educatión de Mexico, S.A. de C.V.
Pearson Education–Japan
Pearson Education Malaysia, Pte. Ltd.

Library of Congress Cataloging-in-Publication Data

Weissman, Jerry.

Prsenting to win : the art of telling your story / Jerry Weissman. — Updated and expanded ed.

p. cm.

Includes index.

ISBN 0-13-714417-2 (hardback : alk. paper) 1. Business presentations. I. Title.

HF5718.22.W45 2009

651.7'3—dc22

2008026750

To the cherished memory of
Louis and Rose Weissman,
my beloved father and mother
. . . fruition at last.

Contents

Foreword to the Updated and Expanded Edition

In the five years since the publication of the first edition of *Presenting to Win*, I am proud to say that it has made a significant impact upon readers, selling more than 100,000 copies in 12 languages. By the same token, I am surprised to say that it has not had as great an impact upon the presentation trade. Despite the many gratifying emails, letters, and telephone calls from around the globe praising the book, and despite the continuing stream of clients that take the Power Presentations program upon which the book is based, I've discovered that most presenters, after reading the book or taking the program, nonetheless default to a practice counter to the main theory in its pages.

Simply put, that theory is stated in the subtitle: *The Art of Telling Your Story*. True to its promise, the book offers techniques about that classic art, but it does so for only two-thirds of its total pages. The other third is about graphic design in presentations, yet that aspect is not even mentioned on the cover. The imbalance is intentional.

The reason for this emphasis on the story, which includes sharp audience focus, clear structural flow, strong narrative linkages, persuasive added value, and even specific positive verbiage, is that the story is much more important than the graphics. No audience will react affirmatively to a presentation based on graphics alone. No decisions are made, no products sold, no partnerships forged, no projects approved, and no ships of state are launched based on a slide show. Witness the powerful speeches that move hearts and minds: State of the Union addresses, inaugurals, nominations, eulogies, sermons, commencements, keynotes, and even locker room pep talks. None of them uses slides.

Therefore, *what* presenters say and *how* they say it are of far greater importance than what they *show*. That is why the lion's share of this book is devoted to helping you tell your story, and why I have even written about the delivery of your story . . . your body language, your eye contact, and your voice . . . in a distinctly separate new book: *The Power Presenter: Technique, Style, and Strategy from America's Top Speaking Coach*.

Does this mean that I am recommending that you abandon all slides ye who enter the podium area? Not at all. Microsoft PowerPoint has become the medium of choice from grade school rooms to corporate boardrooms, and far be it from me to advise a sea change as radical as complete rejection. Graphics play several valuable roles: as illustration of key information, as reinforcement of messages, and as prompts for the presenter, so please leverage this powerful tool.

All I ask . . . no, urge . . . you to do is to use PowerPoint properly, by applying the repertory of techniques provided in the other third of this book. The most essential of which is the overarching principle of relegating your graphics to a supporting role, making you, the presenter, the primary focus.

This seemingly simple plea for a shift of emphasis unfortunately has found very few converts. Presentations are still universally defined by and equated with the slides. This is standard operating procedure with every type of presentation, from IPO road shows to private financing, from product launches to industry conferences, from board meetings to sales pitches, and in every sector of business, from information technology to life sciences, from finance to manufacturing, from pharmaceuticals to real estate, and from media to consumer products. In my 20 years as a coach, I have worked in each of these situations, and have seen this focus on the slides repeated ad nauseam.

Why, then, this misguided imbalance? A brief peek back into history will explain.

Presentations originated as a form of communication back in the dark ages in the middle of the 20th century, when small peer groups within companies gathered around a flip chart perched on a rickety easel to exchange ideas. In that setting, the flip chart became the center of attention as a large surface that all the participants could see and share; but it also served to document the ideas that could later be copied and distributed to others who did not attend the session. The flip chart was such a distinct improvement over the impermanence of a blackboard (and its later cousin, the whiteboard) that it quickly became the display medium of choice in business. In its earliest incarnation then, the sheets of the flip chart served two purposes: as a display *during* the meeting and as a record that could be duplicated and disseminated *after* the meeting. This duality can be described as the *Presentation-as-Document Syndrome*.

This first step in the young life of presentations landed squarely on the wrong foot. By combining the two functions, it formalized an essentially imprudent assumption: that both functions served both purposes when, in fact, they served neither; neither fish nor fowl. A display is *not* a document. A display is for *show* (during the presentation), and a document is for *tell* (after the presentation).

This original sin then proceeded to morph and mutate into its current state of worst practices, driven by successive generations of technological advances.

In the 1960s the medium of choice in the presentation trade had only evolved as far as the primitive overhead projector. That clunky machine, used to display transparent Mylar sheets, known as "foils," stepped up from its humble origins in bowling alleys to take its place front and center in the conference rooms, board suites, and hotel meeting facilities of corporate America.

At root, however, the overhead projector was still just another manifestation of the Presentation-as-Document Syndrome. The document function of the foils became the connection to and the salvation of dispersed participants. Anyone who could not attend the live meeting took up what was to become the hue and cry of business: "Send me a copy of your foils."

In the 1980s, the medium of choice advanced to 35mm slides, and the display took on a more professional look. Nevertheless, this new medium was still hampered by the duality factor, which by then had added new aspects to the document function, now implemented by paper prints of the slides. Documents were no longer merely handouts or "leave-behinds." Their usage widened to include "send-aheads," (before the presentation) speaker notes (crib sheets), validating evidence (exceedingly detailed data), or a manual (of biblical proportions) for consistency of messaging across the company's scattered legions.

Having taken on the status of a business mantra, "Send me a copy of your foils" simply shifted to "Send me a copy of your slides." (Except in some companies, such as Intel Corporation, where even today, although all presentations are done on computers, the employees persist in calling their slides "foils.") Presenters, forced to straddle the functionality fence, generated slides that doubled as documents, heavily weighted toward text and numeric charts. The net effect was a glut of dense eye charts that assaulted the audience's sensory intake. Visual aids became visual hindrances.

In the late 1980s, the PC overtook the carousel projector as the medium of choice for the display function, and the floppy diskette became the medium of distribution for the document function. By this time, however, the term "slides" had stuck. Before or after the meeting, it was still "Send me a copy of your slides." The medium had evolved, but the message stayed the same.

In 1990, Microsoft entered the arena with its release of the Windows version of PowerPoint, an aptly named software application that enabled presenters to make their business points with new and powerful graphics capabilities. Still, despite the continuing evolution of distribution technology from diskettes to CDs to Internet transmission, the business mantra persisted: "Send me a copy of your slides." Pressured by the exigencies of business, beleaguered presenters continued to oblige the request by using the same presentation for both display and distribution, both show *and* tell.

In the meantime, PowerPoint succeeded wildly. Within three years of its launch, it became the market leader, a position it enjoys to this day. Each succeeding generation added more and more features and functions, in the process expanding its installed base around the globe, and beyond business into the not-for-profit world, the government, the military, and even into schools.

Throughout it all, the vestigial legacy of the flip chart endured. The Presentation-as-Document Syndrome continued, and still continues to perpetuate its fowl/fish (pun intended) effect on victimized audiences, where neither version serves its intended purpose, and each version is severely compromised by the dual functionality.

If you need a document, create a document and use word processing software. If you need a presentation, create a presentation and use presentation software. Microsoft Office provides Word for documents and PowerPoint for presentations. While both products are bundled in the same suite, they are distinctly separate entities, and *never the twain shall meet*. Use the right tool for the right job.

Follow the correctly balanced role model you see on all television news broadcasts. The newscasters tell the story, while the professional graphics that flit by over their shoulders are simply headlines.

You are the storyteller, not your slides.

What's Past Is Prologue

My first experience with the power of the spoken word came on December 8, 1941, when, as a child, I joined my father and mother at the family Philco radio to hear President Franklin Delano Roosevelt, in the wake of the attack on Pearl Harbor, deliver his stirring Day of Infamy speech. I'll never forget how he concluded, his rich voice reverberating: "With confidence in our armed forces, with the unbounded determination of our people, we will gain the inevitable triumph. So help us God." In that exhilarating moment, Roosevelt's potent words pierced through our dismay, lifted our spirits, and restored our confidence in our nation and in our future.

Later, I learned more about the ability of words to move people's minds in my graduate classes in the Speech and Drama Department at Stanford University, where I studied the works of the great Greek orators. Still later, in my work as a news and public affairs producer for CBS Television in New York, I witnessed the momentous impact of the words of great national leaders, from John F. Kennedy to Martin Luther King, Jr.

But I never fully realized the universal significance of communication until I left the broadcast medium and entered the world of business. The medium of choice in business is the presentation, and I soon discovered the force it can exert: A poor presentation can kill a deal, while a powerful one can make it soar. Early in my business career, I was privileged to work on the Initial Public Offering presentation, known as an IPO road show, for Cisco Systems, and saw, on its first day of trading after the road show, Cisco's valuation increase by over 40 million dollars.

The big *Aha!* for me was the realization that every communication is an IPO. Everyone communicates every day. You do. I do. Every time we do, we can either fail or succeed. My job is to help you succeed in your everyday communications, just as I helped the Cisco IPO, and as I've helped hundreds of corporations like Microsoft and Intel, and thousands of clients who are executives or managers or salespeople just like you. My job is to help you persuade every audience, every time.

The very same principles that propelled Cisco's success reach all the way back to the classical concepts of Aristotle. Those same basics underlie Abraham Lincoln's towering rhetoric that healed a nation torn asunder by civil war. They underlie Sir Winston Churchill's inspiring orations and Franklin Roosevelt's assuring fireside chats that rallied their nations to the victorious defense of the free world. And they underlie Martin Luther King's rousing speeches that spearheaded the civil rights movement.

They also underlie your sales pitch, your presentation to a potential new customer, your bid for financing, your requisition for more resources, your petition for a promotion, your appeal for a raise, your call to action, and your own quest for the big Aha!

They are the principles that will empower you to present to win.

The Wizard of Aaahs

Once upon a time, I was living and working at the opposite end of California from Silicon Valley, in Hollywood. I had spent the first half of my professional life in the world of show business, as a television producer for CBS, as a freelance screenwriter, and as a paperback novelist. I helped create news documentaries, feature films, dramas, and musicals. I had the opportunity to work with some of the most creative minds in the industry, from the legendary Mike Wallace on down. If you know anything about show business, you know that it's filled with peaks and valleys, and I had more than my share of valleys. But I met many interesting people and learned a lot, particularly about the art of telling a story in a clear, convincing manner.

Then, in 1987, I had a conversation with an old friend, Ben Rosen, one of the top venture capitalists in the high-technology world who was then Chairman of the Board of Compaq Computer Corporation. It was a conversation that changed my life.

Ben and I had met at Stanford University, where he was studying for his Master's degree in electrical engineering and I for mine in speech and drama. The engineer and the artist met only because we happened to be competing for the affections of the same girl. Our interest in the girl quickly faded, but our friendship did not. Ben followed my subsequent career in television and was well aware of my interest in the art of communication. As Compaq's Chairman, he was also aware of an issue facing the great computer company: Its CEO, a talented executive named Rod Canion, had never developed a comfortable and effective style for public presentations.

Ben called to offer me a challenge: "Rod has worked on his weakness as a presenter," he explained. "He's even been coached by some of the experts in the field. But it hasn't quite taken hold. Would you be interested in flying out to Houston to teach Rod what you know about communication?"

I was intrigued, but a little reluctant. After all, I didn't know much about the world of business. But Ben closed the deal with an unusual offer: "Compaq has just come out with a line of hot new laptop computers. I've seen that clunker you're still using." (I'd just laboriously drafted my second novel on Compaq's huge old "luggable" computer and had been coveting the sleek, new, expensive

Compaq machines.) "Suppose we swap you one of our new laptops for your services?" he asked. I agreed on the spot.

I met with Canion at his Houston office, and Ben sat in on our session, watching as I taught Rod the basics of communicating a story with clarity and effectiveness. An hour into the program, we took a break, and Ben buttonholed me at the vending machine in the lounge. He was fascinated by what he'd seen. "Jerry," Ben said, with a snap of his fingers, "There's an enormous business opportunity here! I spend all day listening to presentations by CEOs who want me to invest in their businesses. You wouldn't believe how complex and dry most of them are. You ought to move up to Silicon Valley and teach these people some of your storytelling skills. God knows they need your help!"

Naturally I was flattered. But I thought of myself as a television professional, not as a business consultant. "I don't know anything about Silicon Valley or the computer business!" I protested.

Ben pressed me. "That doesn't matter," he insisted. "I'll be able to introduce you to clients; I can show you how to run the business; I'll help in many ways."

Still I demurred, "It's not a good idea to do business with friends." Ben shook his head and dropped the matter, for the moment.

Like all successful people, Ben is successful because he is persistent, and he persisted with me. He talked about the idea, on and off, for six more months, but I was still hesitant. Finally, at Ben's insistence, I agreed to make a pilot trip to Silicon Valley to meet some of his associates. One of them was Andrea Cunningham, a woman who had parlayed her experience as public relations counsel to Steve Jobs at Apple Computer into her own successful national public relations agency, Citigate Cunningham, Inc.

When I got to Andy's office, she was in a fretful state over a presentation she was scheduled to make at a major technology conference. I took a quick look at a very rough outline she had prepared and suggested a simple reordering of her concepts into a more logical sequence. Then I skimmed through the high points of the new outline for her. Andy's frown gave way to a smile, and she said, "You're going to do very well here!"

My reluctance gradually melted away. I agreed to Ben's business proposition, and Power Presentations was born.

The Mission-Critical Presentation

In the first year of my startup, Ben, true to his word, introduced me to many influential people; primary among them were his venture capital colleagues, or VCs, as they are known in the trade. One of them was one of the most powerful men in Silicon Valley, Don Valentine. A founding partner of Sequoia Capital, one of the premier venture capital firms, Don had been one of the original investors in Apple Computer, Atari, Oracle, and Electronic Arts. Don granted me a courtesy interview, listened patiently as I described my services to him, and then said, "We have a company that's about to go public, and we think it'll do fairly well. It has a very esoteric technology that will be difficult to explain to investors. We're planning on pricing the offering in the $13.50 to $15.50 range, but if the IPO road show presentation is any good, we can probably increase that share price by a couple of bucks. I'm going to introduce you to the CEO and ask him to have you help him with his presentation."

The company was Cisco Systems. The CEO was John Morgridge. I helped John develop the presentation that explained the company's complex networking technology. The message got through to the investors. On the day Cisco went public, its stock opened at $18 a share and closed its first day of trading at $22 a share (a then-unheard-of price jump). Cisco quickly became the darling of the investment community and the media. In an interview with the *San Francisco Chronicle*, Don Valentine, speaking in his role as Chairman of Cisco's board, "attributed 'at least $2 to $3' of the increase to Weissman's coaching. John Morgridge, president of Cisco, is somewhat less generous. He gives Weissman 'at least an eighth of a point'—12 1/2 cents per share."

That was more than 500 IPO road shows ago. Among those others, I coached the IPO road shows of corporate luminaries such as Intuit and Yahoo!. During that same time, I also helped another 500 firms, both public and prepublic, to grow their businesses.

Following the IPO, Cisco continued to call on me for their nuts-and-bolts presentations, ranging from *prezos* (as they are called inside Cisco) in their briefing centers given to small groups of potential new customers, to *prezos* in their annual NetWorkers conferences given to massive assemblages of end users. Cisco's then-Vice President of Corporate Marketing, Cate Muther, required that every product manager take my program. At the time, she said, "Jerry's methods of presentation are now part of our culture; they help prepare our managers for industry leadership." Today Cisco and many other high-technology companies continue to enlist my help in coaching their senior executives to communicate persuasively. The business press has called me everything from the man who "knows how to talk to money" (*Fast Company*) to "The Wizard of Aaahs" (*Forbes*).

The IPO road show is probably the most mission-critical presentation any businessperson will ever make. Succeeding in an IPO road show is the ultimate example of winning over the toughest crowd. The investors are both demanding and knowledgeable, the stakes are high, and a swing of one dollar in the share price of an offering translates into millions of dollars. But if you extend the logic a bit further, every crowd you encounter in business can be your toughest crowd, and every presentation you ever give is mission-critical. Every presentation is a stepping-stone on the path to ultimate success. If any one presentation fails, there may be no tomorrow. You never get a second chance to make a first impression.

You never get a second chance to make a first impression.

Therefore, in my work with my clients, I treat every presentation as if it were as mission-critical as an IPO road show. I use this approach whether I am working to develop a private financing pitch, a product launch, a keynote speech, a panel appearance, an analysts' call, a shareholders' meeting, or a budget approval presentation. You can extend my array with your own presentation situations, be they external or internal, be they for an important contract, a major alliance, or a big sale.

Every business presentation has one common goal: the all-important art of persuasion . . . an art with literally dozens of applications for which *everyone* in business must be prepared. Persuasion is the classic challenge of sounding the clarion call to action, of getting your target audience to the experience known as *Aha!*

Persuasion is the classic challenge of sounding the clarion call to action, of getting your target audience to the experience known as Aha!

In a cartoon, *Aha!* would be represented by the image of a light bulb clicking on above your audience's heads. It's that satisfying moment of understanding and agreement that occurs when an idea from one person's mind has been successfully communicated into another's. This process is a mystery as old as language itself and almost as profound as love; the ability of humans, using only words and symbols, to understand one another and find common ground in an idea, a plan, a dream.

Maybe you've enjoyed moments like this in your past experiences as a presenter, speaker, salesperson, or communicator . . . moments when you saw the light bulb go on, as eyes made contact, smiles spread, and heads nodded. *Aha!* is the moment when you know your audience is ready to march to your beat.

I've written this book to share the persuasive techniques and tools that I provide to my clients. You can use them in your business on a daily basis. The presentation principles my clients have used to attract their billions of investor dollars can work for you, too.

The Art of Telling Your Story

Don Valentine of Sequoia Capital, the legendary venture capitalist who introduced me to Cisco, sits through thousands of presentations every year, most of them made by shrewd entrepreneurs in search of funding for their new business concepts. Don is continually shocked by the failure of most of these presentations to communicate effectively and persuasively.

He once summed it up to me this way: "Jerry, the problem is that nobody knows how to tell a story. And what's worse, nobody *knows* that they don't know how to tell a story!"

The overwhelming majority of business presentations merely serve to convey data, not to persuade.

This problem is multiplied and compounded 30 million times a day, a figure which, according to recent estimates, is how many PowerPoint presentations are made every business day. Presentation audiences, from the Midas-like Don Valentines to overbooked executives sitting through run-of-the-mill staff meetings, are constantly and relentlessly besieged with torrents of excessive words and slides.

Why? Why wouldn't every presenter, seeking that clarion call to action, be, as the U.S. Army urges, all that he or she can be? The reason is that the overwhelming majority of business presentations merely serve to convey data, not to persuade.

When I moved from the world of television to the world of business, I saw immediately that the problem in those massive transmissions of information down one-way streets to passive audiences was not at all communication . . . with the emphasis on the *co-* . . . they were one-way streets that ground to a halt at a dead end.

In the television medium, ideas and images are also broadcast in one direction over the air, cable, or satellite, but there is a return loop, a feedback, an interaction that comes barreling back at the broadcaster in the form of ratings, critics, sponsors, letters, telephone calls, emails, and sometimes even regulatory legislation.

In the Medium that Marshall McLuhan analyzed in his classic book, *The Medium is the Massage* (yes, that's correct, he called it "Massage"), when the message is not clear, and when the audience's satisfaction is not manifest, the foregone conclusion is sudden death: The television series is canceled.

In business, when the point is not crystal clear, and when the benefit to the audience is not vividly evident, the investment is declined, the sale is not made, the approval is not granted; the presentation fails.

In the Power Presentations programs and in this book, you'll find the media sensibility applied to the business community . . . a set of prescriptive techniques and services that will enable presenters like you to achieve their clarion call to action with their audiences, to get them to *Aha!*

A New Approach to Presentations

The techniques you'll be learning in these pages are a blend of new and old concepts from a broad array of disciplines.

When creating television public affairs programs, I had to wade through hours of archival and new film footage and reels of videotape, rifle through massive reports, sort through stacks of interviews, and boil it all down to a clear 28-minute-and-40-second program that would capture . . . and hold . . . the audience's attention. I've netted these story-development methods into a simple set of techniques and forms for businesspeople like you. Most professional writers use these same techniques to tap into natural creative processes that every human being possesses.

In television, I worked in multimillion-dollar control rooms equipped with electronic character generators, vast color palettes, chroma key insertion, and computer-driven on-screen animation. Most of these capabilities are now readily available in Microsoft's PowerPoint software, installed in over 250 million computers. Unfortunately, as anyone who has seen a recent business presentation can attest, most presenters apply these powerful functions with all the subtlety of an MTV video. Instead, they should be following Ludwig Mies van der Rohe's surgically appropriate advice: *Less Is More*. You'll find a simple set of guidelines to help you apply Mies' principle to create visual support for your presentations, to help you design your numeric and text charts so that, rather than overwhelm, confuse, and distract your audience, they enhance and clarify your persuasive message.

In television, I directed both film and video cameras and then spliced and arranged their output into a compelling story. In doing so, I employed the professional practices of cinematography and editing to tell a story and create an impact on the audience. I've translated these sophisticated methodologies into a simple set of continuity techniques . . . that you can readily implement with PowerPoint . . . to help you tell your story.

I've also drawn many of my communication and persuasion techniques from classical sources, such as the writings of Aristotle. (Please don't let the word "classical" intimidate you. A wise person once defined a classic simply as something that endures because it works. You'll rediscover the truth of that definition in these pages.) In ancient times, rhetoric was considered the greatest of the liberal arts; what the philosophers of old called rhetoric is, in fact, what we refer to today as great storytelling. As you read, you'll come to see the relevance of Aristotle's concepts to all the types of stories you need to tell in business . . . storytelling that will persuade your audience to respond to your call to action.

Other methods are based on established knowledge, as well as recent discoveries about the human mind. These scientific findings that detail how *all* brains and eyes absorb information relate directly to how *every* audience reacts to *any* data input.

This combination of traditional and advanced techniques for communication and persuasion will provide what I'm confident you'll find to be a unique and effective system that will help you present to win.

You'll notice that I give frequent and significant emphasis to the word "story," which is intentional. In this book, as in my programs and seminars, the focus is first and foremost on helping you define the elements of your story and the story of your business. The traditional presentation skills . . . body language, gestures, voice production, eye contact, and answering questions from the audience . . . are also important. They are so important that I've covered those subjects in detail in two separate books: *The Power Presenter*, which shows you how to develop your delivery skills, and *In the Line of Fire*, which shows you how to handle tough questions. But in this book, the main focus is on defining your story.

Many clients, when first meeting me, say, "Oh, I don't need any help with my story. Just show me what to do with my hands while I speak. And just show me how to keep myself from saying 'Ummm.'" I tell them that I will address those factors, but only after we've organized their story. There are two very good reasons for this.

First, there are no quick fixes; it takes a considerable amount of time and effort to develop a natural delivery style, unique and yet comfortable to each and every person. (You will, however, find a checklist for the physical presentation environment in Appendix A, "Tools of the Trade.")

Second, and more important, *getting your story straight* is the critical factor in making your presentation powerful. In fact, when your story is right, it serves as a foundation for your delivery skills. The reverse is never true. You may be the most polished speaker on Earth, but if your story isn't sharply focused, your message will fail.

When your story is right, it serves as a foundation for your delivery skills. The reverse is never true.

Let me share an illustrative anecdote:

In 1991, I got a phone call from the public relations people at Microsoft, regular clients of mine. "We have a young executive here named Jeff Raikes," they explained. "He's scheduled to make a presentation about a new product of ours, and we wonder if you could help him prepare. It's called Windows for Pen Computers, the newest member of the Windows family of products."

"Fine," I replied, "Let's book a three-day program for Jeff."

There was a slight pause on the other end of the line. "Well, we're very pressed for time. Jeff has only one day. But it's really important that his delivery skills get polished. He's very smart and knows his stuff, but he just isn't comfortable. Can you make a difference in one day's time?"

"I'll try," I said.

What happened next is revealing. As requested, I spent one day with Jeff. But we had *no* time to work on Jeff's voice or body language. Instead, we used our time to develop a cohesive focus of Windows for Pen Computing: about what this remarkable new software product was designed to do, about the markets Microsoft hoped to serve with it, the history behind its development, the benefits it offered computer users. In short, we created the *story* Jeff had to tell.

I helped Jeff make some decisions about his story: which elements were most relevant and compelling to his audience; which technical details were necessary; and, equally important, which ones were superfluous. Then I helped Jeff organize his presentation so that the key ideas would flow naturally from beginning to end. By the end of the day, we'd worked through the entire story. Jeff was not only in command of the material, but also comfortable about delivering it.

The results? Jeff's presentation went over phenomenally well. Afterward, the public relations people at Microsoft, who knew I'd coached Jeff, praised me for the job I'd done on his delivery skills, though in fact we'd never discussed those at all. *Simply getting the story right helped to transform a hesitant and uncertain speaker into a dynamic and confident one.* Jeff Raikes went on to rise through the ranks to become the president of Microsoft's Business Division and one of the company's most prominent and effective spokesmen. After 27 years, Jeff left to head the Bill and Melinda Gates Foundation, Bill Gates' charitable organization. One of Jeff's last projects at Microsoft was the Tablet PC, the evolved 21st-century version of the technology that began as Windows for Pen Computing.

The lesson: All the vocal dynamics and animated body language in the world can't improve a confusing story, while a clear and concise story can give a presenter the clarity of mind to present with poise.

A clear and concise story can give a presenter the clarity of mind to present with poise.

The Psychological Sell

I've described the classic art of persuasion as getting your audience to *Aha!* To be truly effective, however, one *Aha!* is not enough. The Power Presentation is a continuous series of end-to-end *Aha!*s.

Making presentations is very much like massage therapy. The good massage therapist never takes his or her hands off your body. In the same way, the good presenter never lets go of the audience's minds. The good presenter grabs their minds at the beginning of the presentation; navigates them through all the various parts, themes, and ideas, *never letting go*; and then deposits them at the call to action.

The good presenter grabs their minds at the beginning of the presentation; navigates them through all the various parts, themes, and ideas, never letting go; *and then deposits them at the call to action.*

Notice the verbs in this analogy describing the work of a skilled presenter: *grab*, *navigate*, and *deposit*. All three can be reduced to a least-common-denominator verb: manage. People are the deciding factor in business decisions, and management is the number-one factor in investment decisions. The good presenter is one who effectively *manages* the minds of the audience. Therefore, the subliminal perception of the effective presentation is *Effective Management*.

Of course, no one is ever going to conclude explicitly that a good presenter is an effective manager, a skilled executive, an excellent director, or a superb CEO. That's a bit of a stretch. But the converse proves the point. Unconsciously, the audience makes assumptions. If they are subjected to a presentation whose point is unclear, they will be resistant to responding to the call to action.

Influential investors from Warren Buffett to Peter Lynch subscribe to the commonly held principle of investing only in businesses they understand.

When your story is not clear, when it's fragmented or overly complex, the audience has to work hard to make sense of it. Eventually, this hard work begins to produce first resistance, then irritation, and then loss of confidence.

A book by Steve Krug about Web design has little to do with developing your story, but its title states our point succinctly: *Don't Make Me Think*.

The effective presenter makes it easy for the audience to grasp ideas without having to work. The effective presentation story leads the audience to an irrefutable conclusion. The journey gives the audience a psychological comfort level that makes it easy for them to say "yes" to whatever the presenter is proposing. Presenting, therefore, is essentially selling.

Of course, one can never minimize the importance of having solid factual evidence that validates your business premise. A well-honed presentation is no substitute for a well-conceived business plan, just as a commanding speaking style is no substitute for personal integrity. You must have the steak as well as the sizzle. Yet when two companies or two individuals of equal strength are competing, the winner is likely to be the one who tells the story more persuasively.

In the end, the most subtle impact of a clear and compelling presentation is perhaps the most powerful effect of all: The person who is able to tell an effective business story is perceived as being in command, and deserves the confidence of others. When you are in command of your story, you are in command of the room. Your audience will follow where you lead, and so will money, influence, power, and success.

The person who is able to tell an effective business story is perceived as being in command, and deserves the confidence of others.

This is the core message and value of this book to you, no matter what role or level in business you currently occupy. Perhaps one day you'll go public with a company of your own, and then I hope the techniques you'll learn here will help you make millions of dollars. Before that happens, however, you'll have to get past many mission-critical hurdles, because *every* business decision turns on your ability to tell your story. So please use the same persuasive techniques you'll find in these pages in those scores of other stepping-stones along the way.

To help you master them, you'll see how Cisco persuaded investors to provide billions in capital to support a technology so esoteric that, even today, few people really understand it. You'll see how Yahoo! capitalized on the emerging fascination with the Internet by transforming an irreverent brand image into a meaningful investment presentation. You'll also see how Luminous Networks, a telecommunications startup company, was able to raise $80 million in private capital during one of the steepest market declines in history. I helped create Power Presentations for each of these companies. I hope to do the same for you.

This book is about presentations, yes. But it's about much more than that. It's about psychology, about storytelling, about getting every audience to respond to your call to action. It's about presenting to win.

Chapter One

You and Your Audience

Company Examples

- **Network Appliance**
- **Luminous Networks**

The Problem with Presentations

Few human activities are done as often as presentations, and as poorly. One recent estimate has it that 30 million presentations using Microsoft PowerPoint slides are made every day. I'm sure that you've attended more than a few. How many of them were truly memorable, effective, and persuasive? Probably only a handful.

The vast majority of presentations fall prey to what is known as the *Five Cardinal Sins*:

1. **No clear point.** The audience leaves the presentation wondering what it was all about. How many times have you sat *all the way through* a presentation and, at the end, said to yourself, "What was the point?"
2. **No audience benefit.** The presentation fails to show how the audience can benefit from the information presented. How many times have you sat through a presentation and *repeatedly* said to yourself, "So what?"
3. **No clear flow.** The sequence of ideas is so confusing that it leaves the audience behind, unable to follow. How many times have you sat through a presentation and, at some point, said to yourself, "Wait a minute! How did the presenter get *there*?"

4. **Too detailed.** So many facts are presented, including facts that are overly technical or irrelevant, that the main point is obscured. How many times have you sat in on a presentation and, at some point, said to yourself, "What does *that* mean?"
5. **Too long.** The audience loses focus and gets bored before the presentation ends. How many times in your entire professional career have you *ever* heard a presentation that was too *short*?

When presenters commit any of these sins, they are wasting the time, energy, and attention of their audience. What's more, they are thwarting their own objectives.

Each of these Five Sins is quite separate and distinct from the others. Here's an analogy to illustrate:

Suppose you and I were chatting, and I said, "Let me tell you about what I had for dinner last night." My presentation would have a point, wouldn't it? You'd know what I intended to do, and I wouldn't be committing Sin #1.

But why on Earth should you care about what I had for dinner last night? Unless you had said, "Jerry, I'm bored with all the restaurants in the area. Can you recommend a new place?" Then, by telling you about the excellent meal I had at a hot new bistro last night, I would be providing a benefit to you, and I'd avoid Sin #2.

Now, if I told you about my fine meal by starting with the dessert, then I went back to the salad, then jumped forward to the cheese course, then back to the main course, my story would have no flow. If instead I went from soup to nuts, it would have a clear and orderly path, and I'd eliminate Sin #3.

If I described the courses I ate by using the phylum, class, order, genus, and species of every animal and vegetable product in the dinner, it would be far too technical and too detailed. If instead I confined my description to colorful adjectives and simple nouns, I would avoid Sin #4.

Finally, if I took five hours to tell you about a meal that took me only two hours to consume, my presentation would be too long. If instead I did it in five minutes, I'd escape Sin #5.

This analogy may be extreme, but the Five Cardinal Sins are all too real. And although each of the five is unique and independent of the others, they can all be summarized in a least common denominator, a Data Dump: an excessive, meaningless, shapeless outpouring of data without purpose or plan. The inevitable reaction of audiences to a Data Dump is not persuasion, but rather the dreadful effect known as *MEGO: Mine Eyes Glaze Over.*

The inevitable reaction of audiences to a Data Dump is not persuasion, but rather the dreadful effect known as MEGO: Mine Eyes Glaze Over.

Why? Why would any presenter in his or her right mind do that to any audience? Would you do that if you were trying to attract potential clients? Would you do that if you were trying to clinch a sale, raise investment capital, or convince analysts that your company is solid? Hardly.

The objectives of all the preceding presentations are varied, but they all have one factor in common. In every case, you are trying to persuade your audience to do your bidding, to respond to your call to action, whether that means endorsing a proposal, signing a contract, writing a check, or working harder and smarter. The Five Cardinal Sins stand in the way of achieving that goal.

The Power Presentation

Most people in business, including the most successful ones, are too busy *living* their stories to focus on *telling* them. They spend 12 or 14 hours every day working on competitive strategies, product launches, financial analyses, marketing plans, mergers and acquisitions, sales pitches . . . the plethora of vital business details that fill *your* days, too. They live, eat, sleep, breathe, dream, and inhale their businesses. They see every single one of the trees, but not the forest. They rarely have the opportunity, or feel the need, to take several long steps back from the details to visualize the whole and then describe it compellingly.

For most businesspeople, when a situation arises in which they *must* sell their business story, their intense involvement in the minutia often proves to be a hindrance. They mistakenly think that for the audience to understand *anything*, they have to be told *everything*. That's like being asked the time and responding with complete instructions for building a clock.

The remedy is painfully apparent: Focus. Separate the wheat from the chaff. Give the audience only what they need to know.

Persuasion: Getting from Point A to Point B

As social animals, we humans find ourselves called upon to persuade other humans almost every day. Persuasion is one of the crucial skills of life. The persuasive situations you may face will be remarkably varied, each posing its own unique challenges and opportunities.

Yet all presentation situations have one element in common. Whether it's a formal presentation, speech, sales pitch, seminar, jury summation, or pep talk, every communication has as its goal to take the audience from where they are at the start of your presentation, which is *Point A*, and move them to your objective, which is *Point B*. This dynamic shift is persuasion.

Every communication has as its goal to take the audience from where they are at the start of your presentation, which is Point A, and move them to your objective, which is Point B.

Recognizing this truth is the best starting point to learn the art of persuasion. Your presentation may be entertaining, eloquent, or impressive, but that's not your main purpose in offering it. Your main and *only* purpose is to move people to Point B. That's the point! When your point is not clear, you have committed one of the Five Cardinal Sins; when your point, Point B, is readily apparent, you have made your clarion call to action.

Let's take a closer look at what's involved in the challenge of moving an audience from Point A to Point B. In psychological terms, Point A is the inert place where your audience starts: *uninformed*, knowing little about you and your business; *dubious*, skeptical about your business and ready to question your claims; or, in the worst-case scenario, *resistant*, firmly committed to a position contrary to what you're asking them to do.

What you are asking them to do is Point B. The precise nature of your Point B depends on the particular situation you face. To reach Point B, you need to move the *uninformed* audience to *understand*, the *dubious* audience to *believe*, and the *resistant* audience to *act* in a particular way. In fact, *understand*, *believe*, and *act* are not three separate goals, but three stages in reaching a single, cumulative, ultimate goal. After all, the audience will not act as you want them to if they don't first *understand* your story and *believe* the message it conveys.

When I coach the executives of a company to prepare for their IPO road show, the audience for whom they're preparing will be composed of prospective investors. Point B is the moment at which those investors are willing to sell some of their holdings in Intel or Microsoft and invest those valuable assets in shares of the fledging company.

Dan Warmenhoven, the CEO of Network Appliance at the time the company went public, began his IPO road show with this opening statement: "What's in a name? What's an appliance? A toaster is an appliance. It does one thing and one thing well: It toasts bread. Managing data on networks is complicated. Until now, data has been managed by devices that do many things, not all of them well. Our company makes a product called a file server. A file server does one thing and does it well: It manages data on networks."

If Dan had stopped there, his investor audience would have understood what his company did. But I coached Dan to go beyond that description to add: "When you think of the explosive growth of data in networks, you can see that our file servers are positioned to be a vital part of that growth, and Network Appliance is positioned to grow as a company. *We invite you to join us in that growth.*"

That last sentence is the call to action. Notice that Dan did not ask the investor audience to buy his stock. That would have been presumptuous and unnecessary; after all, their very job titles included the word "invest." But the additional sentences gave Dan the opportunity to lead his audience from Point A to Point B. A synonym for "lead" is "manage." The subliminal perception, then, is *Effective Management*.

Start with the Objective in Sight

Point B, then, is the endgame of every presentation: its goal. The only sure way to create a successful presentation is to begin with the goal in mind.

This is an age-old concept. Aristotle called it *teleology:* the study of matters with their end or purpose in mind. Today's business gurus market the same idea. In Stephen R. Covey's *The Seven Habits of Highly Effective People,* he stresses the importance of starting with the objective in sight. Aristotle in modern terms.

Few executives study Aristotle these days, but Covey has been read by millions, to say nothing of the countless others who've heard about his ideas from friends and colleagues. Nonetheless, this crucial concept of starting with the goal in mind hasn't penetrated our thinking about presentations. Think of the many times when, after sitting through an entire presentation, you asked yourself,

"What was the point?" One of the Five Cardinal Sins. The missing point is Point B: the call to action.

Unfortunately, and inexplicably, Point B is missing from all too many presentations.

If you're a sales professional, how can your customer reach the point of making a purchase unless you ask for the sale? If you're a corporate manager, how can the members of your team agree to support your new business initiative unless you tell them unmistakably what you need them to do, and explicitly ask for their help? If you're an ambitious young worker, how can your manager give you the raise or promotion you desire unless you ask for it?

Obvious? Maybe. But it's surprisingly common for businesspeople to forget to focus on Point B when they communicate. If you start your persuasive process with a clear focus on Point B, you'll have a far better chance of ending there, accompanied by your audience. Ask for the order! Call your audience to action! Get to the point! Get to *your* Point B!

Audience Advocacy

For the presenter to succeed in achieving the clarion call to action, the audience must be brought into equal focus with the presenter's objectives. To establish that balance, let me introduce the term *Audience Advocacy*. Mastering Audience Advocacy means learning to view yourself, your company, your story, and your presentation through the eyes of your audience.

Mastering Audience Advocacy means learning to view yourself, your company, your story, and your presentation through the eyes of your audience.

In programs with my clients, I role-play potential investors, prospective customers, or would-be partners. In developing my own program material, I take the point of view of my clients. You must do the same in whatever presentation you are developing. Take your audience's point of view. This is a shift in thinking that requires both knowledge and practice.

Let's refer again to Aristotle, that pioneer in the art of persuasion. In his master work, the *Rhetoric*, Aristotle identified the key elements of persuasion, the most important of which he called, in the Greek of his day, *pathos*. Pathos refers to the persuader's ability to connect with the feelings, desires, wishes, fears, and passions of the audience. The English words we use today reveal connections with the ancient Greek root of pathos. Think of "empathy" and "sympathy," for example. Aristotle wrote: " . . . persuasion may come through the hearers when the speech stirs their emotions. Our judgments when we are pleased and friendly are not the same as when we are pained and hostile."[1]

The question is: How can you *communicate* so that your audience will be pleased, friendly, and ready to act on your Point B? My experience, and that of thousands of my clients, suggests that the best method is Audience Advocacy. Everything you say and do in your presentation must serve the needs of your audience.

It's a simple concept, yet profoundly important. If Audience Advocacy guides your every decision in preparing your presentation, you'll be effective and persuasive.

Shift the Focus from Features to Benefits

One way to understand the concept of Audience Advocacy more fully is via one of the classic rules of advertising and sales, still emphasized in those professions today because it is so fundamental . . . the distinction between Features and Benefits. This distinction is vitally important whenever you're called upon to sell your story. In fact, when you shift the focus of any presentation from Features to Benefits, you heighten the chances of winning converts to your cause.

A Feature is a fact or quality about you or your company, the products you sell, or the idea you're advocating. By contrast, a Benefit is how that fact or quality will help your audience. When you seek to persuade, it's never enough to present the Features of what you're selling; every Feature must always be translated into a Benefit. Whereas a Feature may be irrelevant to the needs or interests of your audience, a Benefit, by definition, is always relevant. Without Benefits, you have no Audience Advocacy. For people to act on anything, they must have a *reason* to act, and the reason must be *theirs*, not yours.

For people to act on anything, they must have a reason *to act, and it must be* their *reason, not yours.*

The same principle applies to any persuasive challenge you face. Features are of interest only to the persuader; Benefits are of interest to the audience. Go with Benefits every time.

1 Aristotle. *Rhetoric*. Translated by W. Rhys Roberts (New York: Random House, 1954).

Understand the Needs of Your Audience

You can create an effective presentation only if you know your audience: what they're interested in, what they care about, the problems they face, the biases they hold, the dreams they cherish. This means doing your homework. If you're in sales, for example, it's imperative that you take the time to get to know your customers: how and why they could use your product, their financial constraints, their competitive issues, and how your product can help them achieve their personal or professional goals. And while you need to understand them as representatives of the marketplace or of a client company, you also need to understand them as human beings. What are their biggest headaches, fears, worries, aspirations, needs, loves, and hates? How can what you have to offer serve them?

At times, your interests and those of your audience are bound to diverge, which creates the potential for conflict and frustration. You may dearly desire that raise, that lucrative sales contract, or that crucial loan or investment needed to keep your business afloat. Inevitably, your audience members will have their own motivations and issues that differ from yours. The art of persuasion must be *balanced* by Audience Advocacy: convincing your audience that what you want will serve *their* interests, too.

Alex Naqvi was the CEO of Luminous Networks, a private Silicon Valley company that has since been acquired by Adtran. Luminous' technology provided optical Ethernet solutions that enabled the giant telecommunications carriers to deliver, on a single platform, a combination of Internet traffic, interactive and broadcast video, and voice services. Although a veteran in the industry, Alex usually found the telecoms, as they are known in the trade, a crowd that is an especially tough sell. But Alex learned to recognize, understand, and respond to the interests and feelings of those audiences. Alex explained:

> *Our new technology makes it possible for telecoms to deliver better Internet service more economically than ever before. We thought that using Luminous Networks would be a no-brainer for any telecom manager.*
>
> *Unfortunately, we weren't considering the point of view of our audience. I'm thinking of one potential customer in particular: a big, important telecom with a long history in the industry. Many of the managers we were hoping to sell our services to had been with the firm for 20 years. They were conservative and maybe a little afraid of the new and the unknown . . . both of which Luminous represented.*

In our early days, we didn't understand how to reach out to an audience like that. We went in with a rather cocky attitude, talking about how our technology was "a radical paradigm shift for the new century." We described its advantages in a way that implied that anyone who didn't get it was probably kind of dumb.

Looking back, it's easy to see our mistake. We were alienating the very people we needed to win over. No wonder they didn't want to buy from us.

In time, we learned to soften our presentations. We started describing our technology not as a radical shift, but as a natural evolution from the current technology. We learned to send the message: "You're not dumb. The technology you now have in place was perfectly appropriate for its day. But now the world has changed, and Luminous is ready with the next-generation technology you and your customers need." As you can imagine, our sales results are a lot better with this approach!

It's funny: As engineers, we tend to look at the challenge of selling our story as a lifeless, logical proposition. We forget the human factor. The message must be honed to address those human motivations. We forget that it's living people we are selling to!

Getting *Aha!*s

Let's review what we've discussed so far.

When stories are complicated, when continuity is absent, when the audience is overwhelmed, and when the presenter doesn't establish a clear bond with the audience, presentations fail. The MEGO syndrome sets in. As a result, *no business gets done*: The investment isn't made, the deal isn't consummated, the sale doesn't happen.

Your goal, of course, is just the opposite: to get your audience figuratively or literally to ask, "Where do I sign?" That is the essence of persuasion.

Persuasion is the art of moving your audience from Point A, a place of ignorance, indifference, or even hostility, toward your goal . . . navigating them through an unbroken series of *Aha!*s . . . to Point B, a place where they will act as your investors, customers, partners, or advocates, ready to march to your drum.

You can move your audiences along this path only when you follow the principles of Audience Advocacy: to place *their* needs at the heart of *your* presentation. The central expression of Audience Advocacy is presenting Benefits rather than Features.

Few communicators achieve the sheer exhilaration of end-to-end *Aha!*s. But most communicators can come a lot closer than they usually do . . . as you will when you apply the Power Presentations techniques in this book.

Chapter Two

The Power of the WIIFY

Company Examples

- **Brooktree**
- **Netflix**
- **Luminous Networks**

What's In It For *You*?

The key building block for Audience Advocacy, and a way to focus on benefits rather than features, is to constantly ask the key question: What's in it for *you*? It's based on the more common axiom "What's in it for *me*?," but we've shifted the ultimate word to "you" deliberately, to shift the focus from you to your audience. This shift emphasizes the ultimate need for all communicators to be focused outward, on the needs of their audience ("you"), rather than on their own needs ("me"). This is the essence of Audience Advocacy.

In referring to this key question, let's use the acronym WIIFY (pronounced "whiffy"). By constantly seeking the WIIFY in any persuasive situation, you can ensure that your presentation stays focused on what matters most: getting your audience to move from Point A to Point B, because you've given them a very good reason to make that move.

The WIIFY is the benefit to the specific audience in your persuasive situation. Usually there will be one overarching, grand WIIFY that unites the entire presentation and is at the heart of your persuasive case.

For example, when an entrepreneurial CEO and his or her management team launch an IPO road show for potential investors, the WIIFY is, "If you invest in our company, you'll enjoy an excellent return on your money!"

On the other hand, when a corporate headhunter makes a job offer to a sought-after young recruit, the WIIFY is, "If you join our firm, you'll be starting an incredible career with great pay, fascinating challenges, and the prospect of some day becoming the company president!"

When a partner in a marketing consulting firm makes a new business proposal to the Chief Operating Officer (COO) of a Fortune 500 company, the WIIFY might be, "If you retain us, the expertise we'll provide will improve your promotional plans, which can increase your market share and boost your profits!"

WIIFY Triggers

In addition to these grand WIIFYs, there will usually be many smaller WIIFYs . . . more-specific, but still significant, audience benefits that give meaning to each element in your presentation. In fact, every element in your persuasive presentation must be clearly linked to a WIIFY.

There are six key phrases that can trigger a WIIFY. They are designed to remind presenters about the necessity of linking every element of their presentation to a clear audience benefit, or, a WIIFY. When I coach my clients' presentations, if I hear an idea, fact, story, or detail *without* a clear audience benefit, I interrupt to call out one of these WIIFY triggers:

1. "This is important to you because . . . ." (The presenter fills in the blank with a WIIFY.)
2. "What does this mean to you?" (The presenter explains with a WIIFY.)
3. "Why am I telling you this?" (The presenter explains.)
4. "Who cares?" ("You should care, because . . . .")
5. "So what?" ("Here's what . . . .")
6. "And . . . ?" ("Here's the WIIFY . . . .")

Get to know these WIIFY triggers. Use them on yourself the next time you're preparing a presentation, as reminders to link every element to a WIIFY. (You might want to copy this list from Appendix B, "Presentation Checklists," and tack it on the wall for continual inspiration.) When you work on a presentation as part of a team, use these triggers on your colleagues, and encourage them to return the favor. By the tenth time you pull a WIIFY trigger, you may catch a nasty look or two, but the quality of the resulting presentation will make it all worthwhile.

Here's an example of how important it is to constantly translate your ideas into WIIFY terms. Jim Bixby was the CEO of Brooktree, a company that made and sold custom-designed integrated circuits used by electronics manufacturers. (The company was later acquired by Conexant Systems, Inc.) In preparation for Brooktree's IPO, Jim rehearsed his road show with me. I role-played a money manager at Fidelity, considering whether our mutual fund might invest in Brooktree. During the product portion of Jim's presentation, he held up a large, thick manual and said, "This is our product catalog. No other company in the industry has as many products in its catalog as we do in ours."

Jim set down the catalog and was about to move on to the next topic when I raised my hand and fired off a WIIFY trigger. "Time out!" I said. "You say you have the biggest product catalog. Why should I care about the size of your catalog?"

With barely a pause, Jim raised the catalog again and replied, "With this depth of product, we protect our revenue stream against cyclical variations."

The lights went on. This was an immensely important factor in the company's financial strength, yet one that could easily have passed unnoticed, simply because Jim had forgotten to ask himself, "What's the WIIFY?" Always find and state your WIIFY!

Always find and state your WIIFY!

In any presentation, before you make *any* statement about yourself, your company, your story, or the products or services you offer, stop and ask yourself: "What's the WIIFY? What benefit does this offer my listener?" If there is none, it's a detail that may be of interest to you and your colleagues (a feature), but one that has no significance to your audience. But if there is a benefit, be sure you explain it, clearly, explicitly, and with emphasis, just as Jim did when I pulled the WIIFY trigger on him.

At this point, you may want to protest, "Wait a minute. My audiences aren't stupid. They can figure out the benefits of whatever I mention. They might even feel insulted if I spell it all out for them!"

This is not necessarily true. Remember the Five Cardinal Sins. One is lack of a clear benefit. An essential truth about Audience Advocacy is that most businesspeople today are overloaded with information, with commitments, with responsibilities. When you make your presentation, you *may* have your audience's undivided attention . . . but not necessarily. Even if it takes them just a few seconds to connect the dots between the feature you describe and the implied benefit, by the time they catch up, you will have moved on to your next point, and they probably won't have time to absorb the benefit . . . or the next point. You'll have lost your audience, perhaps permanently.

By stating the WIIFY, you seize an opportunity. Although your audience members are eminently capable of realizing the WIIFY on their own, when you state it for them, you lead them toward a conclusion . . . your Point B. In doing this, you manage their minds, you persuade them, and you instill confidence in your story, your presentation, and yourself. Plus, you accomplish something else. The audience may have just gotten to the *Aha!* themselves, a moment before you stated the WIIFY. By articulating it, you win their agreement. They react with nods, thinking to themselves, "Of course! I've never heard it put so succinctly and clearly!" *Effective Management.*

This is a variation on the Features/Benefits distinction. When presenting to potential investors, a CEO may explain the best features of a leading product: "We've built a better mousetrap." But it's not the quality of the mousetrap in itself that the investors care about; it's the size of the market. The effective CEO presenter will then promptly move on to state the benefit to investors: " . . . and the world is beating a path to our doorstep." There is a huge market for mousetraps. When the WIIFY is right, everybody wins.

In fact, the power of the WIIFY even applies in our personal lives. Consider this example:

Debbie runs a small but growing catering business. In the past, she has managed to keep most of her weekends largely free of work, which her husband, Rich, thoroughly appreciates. Now, however, she's received the proverbial "offer she can't refuse": a request to cater a series of receptions at the local art museum that will keep her busy on weekends throughout the fall and winter. It will be quite lucrative as well as prestigious, but Debbie has to convince Rich to support her in this endeavor. Over dinner one evening, Debbie paints an eloquent word picture of how catering the receptions will put her company on the map, *but she doesn't tell Rich how he'll benefit.*

To win Rich's support, Debbie should say something like this: "This contract could boost my profits from the catering business next year by over 50 percent. It'll be enough to let me hire an assistant manager who can run the business for three weeks next summer . . . while we take that European tour we've always dreamed about."

In this example, Debbie had to do more than simply reframe the idea to make the WIIFY clear. She also had to adjust her plans so that Rich will receive a definite personal benefit. One of the advantages of crafting a well-conceived WIIFY is that if you haven't previously shaped your proposition to be a true win/win deal with benefits for everybody, then presenting the WIIFY will impel you to do just that. Improving your presentation can also help to improve the underlying substance.

The Duchess of Windsor famously once said: "You can never be too thin or too rich." I amend her adage with " . . . or offer too many WIIFYs."

The Duchess of Windsor famously once said: "You can never be too thin or too rich." I amend her adage with " . . . or offer too many WIIFYs."

The Danger of the Incorrect "You"

One seemingly obvious aspect of the WIIFY principle that proves to be a stumbling block for many businesspeople is the danger of the incorrect "you." Let me demonstrate:

One of my clients (let's call him Mark) was the CEO of a company that manufactured dental instruments, wonderful tools of exceptional quality and precision to perform root canal procedures. Mark's prior experience had been as a top salesman for another dental instrument company. Now, as CEO, Mark was preparing to take his new company public. I coached him through a rehearsal of his IPO road show by role-playing, as I usually do, a high-powered fund manager at Fidelity.

Mark eloquently presented his company's strengths, focusing in particular on the high quality of their products. As an example, he described the special features of the new dental instrument his company had developed. Mark held up the actual instrument, looked at me, and said, "With this instrument, you can do better root canal procedures, more quickly and with less pain."

I stopped him. "That's fine," I said. "But I'm an investor, remember? I don't do root canals!"

"Hmm," said Mark. He smiled, thought for a moment, then held up the instrument again, and said, "So you can see that the tens of thousands of endodontists across this country and thousands more around the world who want to do better root canals need instruments like this one, and they'll have to buy them from us!"

Now that's the correct you!

You can see Mark's problem: In trying to formulate the WIIFY, he'd lost sight of his audience, the "you" of the question "What's in it for you?" Instead, he devised a WIIFY that referred to the ultimate end user of his product, the endodontists, to whom he'd previously been selling. To hone his appeal to the investors who were now his audience, it was necessary for Mark to carefully focus on their concerns, which related to the size of the market for his instruments.

Can you get away with the incorrect you? Will your audience be able to translate the benefit to another party into terms that are meaningful to them? Of course they can. But if they do, they will have to make a split-second interpolation to adjust to the correct you. During that interval, they may stop listening to you and start thinking.

Don't make them think!

Consider those words as a guideline for Audience Advocacy. Make it easy for your audience to follow, and your audience *will* follow your lead.

Make it easy for your audience to follow, and your audience will *follow your lead.*

What if the audience did make the leap themselves, translating the WIIFY into terms that are meaningful to them? In Mark's case, he would still be missing a golden opportunity to manage his audience's mind to Point B.

This problem of the incorrect you is a surprisingly common one. Many of us in business have to sell ourselves and our stories to multiple constituencies, each with different biases, goals, preferences, interests, and needs. It's easy to lose sight of today's audience and address another audience's WIIFY.

Here's another example of the problem of the incorrect you:

Reed Hastings is the CEO of Netflix, an online DVD subscription company that went public in the spring of 2002. I had worked with Reed back in 1996, when he headed another company called Pure/Atria Software. At that time, I taught Reed, as I do all my clients, the subtle but important difference of addressing the correct you. Nonetheless, when Reed emailed me the draft of his road show for Netflix, one of the first slides in the presentation described his core business as shown in Figure 2.1.

Figure 2.1 *Netflix business description slide, first draft.*

When Reed arrived for our coaching session, I assumed my usual role as a potential investor in Netflix's stock offering. I said, "Reed, this presentation makes me really eager to sign up and become a loyal subscriber of Netflix . . . but you didn't come here today to get me to subscribe. I can sign up on the Internet. Treat me as an investor."

Reed smiled and said, "What do you suggest?"

On the computer, I revised Reed's slide to read as it does in Figure 2.2.

Figure 2.2 *Netflix business description slide, second draft.*

Suddenly, the entire frame of reference changed from the attractiveness of Netflix's consumer offering to how large the market opportunity was . . . a much more important consideration for Reed's investor audience.

Reed smiled broadly and said, "How about *tens* of millions of movie lovers?"

"Great!" I concurred. "How about 'tens of millions in the U.S. *alone*'?"

Reed accepted the revision, polished his presentation, and then left to begin his IPO road show. One month later, when Netflix went public, they offered 5.5 million shares for sale. They received orders for 50 million shares . . . oversubscribed by nearly 10 times.

On their own, the members of the investor audience could have readily deduced that "Rent All the DVDs You Want" really referred to the many millions of potential Netflix customers, but then the audience would have been doing the math for themselves. By providing the logic for them, Reed led them to a conclusion and, in doing so, built their confidence. Reed seized his opportunity.

Never take the "you" in the WIIFY for granted. It's always necessary to give deliberate thought to who your audience is and what they want. If your WIIFY is designed for the wrong ears, it can fall flat.

Also note this: The problem of the incorrect you is a major reason to resist the temptation to create a generic presentation about yourself, your company, or your products. The generic presentation, or "the company pitch," as it is frequently called, assumes that the same presentation can be used with few or no changes for a variety of audiences. However, the same story that excites and inspires your own employees may bore your customers and actually alienate and anger your suppliers, or vice versa. The same story that persuades technical customers to buy your product may confound your potential investors.

A perfect case in point comes from Alex Naqvi, the former CEO of Luminous Networks, whom we met in the previous chapter. Luminous, which had started in business in 1998, planned eventually to go public, but given the challenging market conditions in 2001, Alex and his team decided to take their show on the private, rather than the public, road to seek additional financing.

> *Before we presented to the investors, we also did due diligence on them and their professional backgrounds. If they were from the technology industry or had worked for one of the carriers, I would tell the story differently; I'd use technology buzzwords that I knew they would understand. But if the investors had formerly been investment bankers, I'd explain our business differently. The key is that everyone in the audience should be able to relate to what I'm talking about.*

We did a total of about 60 presentations. It was a very tough environment, a poor financial market. But in the end, the presentation helped us raise the money we needed . . . 80 million dollars. When I tell people about it, they don't believe that we were able to achieve that given the tough climate we were operating in.

Although the title of this chapter is "The Power of the WIIFY," it could also be considered "The Power of 'You.'" "You" generates innumerable potent benefits, all of them targeted at persuading your audience. "You" is so powerful that I have developed a simple rule I apply to everything I write: books, articles, letters, and emails. Before dispatching any document, I do one final review to see if I can insert additional instances of "you." In this chapter alone, "you" has occurred 76 times thus far, not counting its variations of "your," "yours," and "yourself."

For example, if I met someone at a business function, I might send that person a follow-up email:

It was good that we met at the conference. I look forward to future meetings.

With the "you" rule applied, it would read as follows:

It was good to have met you at the conference. I look forward to meeting you again in the future.

Try the "you" rule yourself. It will personalize the tone and heighten the impact of any book, article, letter, or email you write.

Chapter Three

Getting Creative: The Expansive Art of Brainstorming

Company Example

- **Adobe Systems**

The Data Dump

As you've seen, a critical component of crafting a winning presentation is that, first and foremost, you must get your story right. Although a strong speaking voice, appropriate gestures, and skilled answers to challenging questions are important factors, none of them will yield a really powerful presentation unless your story is clear and leads your audience directly where you want them to go: your Point B.

Creating your presentation begins with the development of your story. Here is one of the first places where traditional methods of creating a presentation can go wrong.

Remember the MEGO syndrome? It strikes when Mine Eyes Glaze Over during a presentation that overflows with too many facts, all poured out without purpose, structure, or logic. When that happens, the presentation degenerates into a Data Dump: a shapeless outpouring of everything the presenter knows about the topic.

All too many businesspeople labor under the mistaken assumption that, for their audience to understand *anything*, they have to be told everything. As a result, they give extensive presentations that amount to nothing more than Data Dumps: "Let's show them the statistics about the growth of the market. Then we've got the results of the last two customer satisfaction surveys. Throw in some excerpts from the press coverage we got after our product launch. Give them the highlights of our executive team's resumes. And don't forget the financial figures . . . the more, the better." This is known as the Frankenstein approach: assembling disparate body parts.

The audiences to these Data Dumps are hapless victims. But sometimes the victims rebel. "And your point is?" and "So what?" are the all-too-common anguished interruptions of audiences besieged and overwhelmed by torrents of excessive words and slides. Those interruptions, however, are made more out of self-protection than rudeness. *The fault, dear Brutus, is not in our stars, but in ourselves.*

Hopefully, you'll never inflict a Data Dump on any of your audiences. But performing one is vital to the success of any presentation. The secret: The Data Dump must be part of your *preparation*, not the presentation. Do it backstage, not in the show itself. (The Greek word "obscene" originally described any theatrical action, such as a murder, that was kept offstage, out of the scene, because it was improper to display such behavior in public. In this sense, you can regard a Data Dump as literally obscene.)

The secret: The Data Dump must be part of your preparation, *not the presentation.*

What you need instead is a proven system to incorporate a thorough Data Dump into the development of your story. Brainstorming is that system. It's a process that encourages free association, creativity, randomness, and openness while helping you consider all the information that may (or may not) belong in your presentation. Later on in the process, you can sort, select, eliminate, add, and organize these raw materials into a form that flows logically and compellingly from Point A to Point B. At the start, the key is not to apply logic to the materials, but simply to get them all out on the table, where they can be examined, evaluated, and sorted. Do the distillation *before* the organization: Focus *before* Flow.

Do the distillation before the organization: Focus before *Flow.*

To understand this approach, it's important to consider the different processes and skills that go into the creative effort of developing a presentation. These are concepts drawn from my shared experiences with professional creative people in the media.

Left Brain Versus Right Brain

Scientists have long been fascinated by how different mental functions are centered in different areas of the human brain. Most of the higher brain activities occur in the cerebrum, which is divided into left and right hemispheres. According to most scientists, the left and right halves of the brain are responsible for different forms of reasoning. The left side controls *logical* functions and is associated with structure, form, sequence, ranking, and order. It tends to operate in a linear, first-one-thing-and-then-the-next fashion. The right side controls *creative* functions and is associated with concepts. It is essentially nonlinear in its operations. The right brain bounces around among concepts, following connections that are impossible to explain logically.

Building a presentation is a *creative* process. That means starting with the right brain.

Here's the problem: Most presenters, when developing their stories, tend to apply a left-brain approach to what is really a right-brain process. They try to jump immediately to a logical, structured, linear end product when their right brain is still caroming in nonlinear mode.

Why? Because businesspeople are results-oriented rather than process-oriented. I'm sure that you, like most businesspeople, are quite process-oriented when it comes to critical matters such as long-term strategy, product design, or problem-solving, but these are all subjects for off-site meetings. *Backstage!*

When it comes to results-oriented tasks, such as developing a presentation to a very important audience (*front and center stage*), you are eager to get there in the shortest distance between two points. You think that a time-consuming process might delay the result. But if you choose a left-brain process while your right brain is free-associating, you'll wind up traversing that seemingly short distance between points over and over and over again. You'll end up spinning your wheels and spilling out a Data Dump.

The solution is timing. It's not a matter of more time; it's about the proper use of time. Get the sequence right: Let the right brain complete its stream-of-consciousness cycle *before* applying the left brain's structure. Focus *before* Flow.

Let the right brain complete its stream-of-consciousness cycle before *applying the left brain's structure.*

A vivid illustration of the distinct difference between right and left brain functioning is spoken language. Speech reflects how the right brain operates in its spontaneity, in its grammatical and syntactical messiness, and in its frequent logical leaps.

Let's illustrate with an excerpt from the live presidential debate between then-Governor George W. Bush of Texas and Vice President Al Gore. The debate, in a town hall format, took place on October 17, 2000, and PBS anchor Jim Lehrer was the moderator. Each candidate was given a chance to respond to questions posed by ordinary citizens. Here is Governor Bush's response to the following question: "How will your tax proposals affect me as a middle-class 34-year-old single person with no dependents?"

> *You're going to get tax relief under my plan. You're not going to be targeted in or targeted out. Everybody who pays taxes is going to get tax relief. If you take care of an elderly in your home, you're going to get the personal exemption increased.*
>
> *I think also what you need to think about is not the immediate, but what about Medicare? You get a plan that will include prescription drugs, a plan that will give you options. Now, I hope people understand that Medicare today is—is—is important, but it doesn't keep up with the new medicines. If you're a Medicare person, on Medicare, you don't get the new procedures. You're stuck in a time warp, in many ways. So it will be a modern Medicare system that trusts you to make a variety of options for you.*
>
> *You're going to live in a peaceful world. It'll be a world of peace, because we're going to have clearer—clear-sighted foreign policy based upon a strong military, and a mission that stands by our friends—a mission that doesn't try to be all things to all people. A judicious use of the military which will help keep the peace.*
>
> *You'll be in a world, hopefully, that's more educated, so it's less likely you'll be harmed in your neighborhood. See, an educated child is one much more likely to be hopeful and optimistic. You'll be in a world in which—fits into my philosophy, you know, the harder work—the harder you work, the more you can keep. It's the American way.*
>
> *Government shouldn't be a heavy hand. That's what the federal government does to you. Should be a helping hand. And tax relief in the proposals I just described should be a good helping hand.*[1]

1 www.c-span.org/campaign2000/transcript/debate_101700.asp

Remember (if you still can) that the original question had to do with how a 34-year-old single person with no dependents would be affected by the candidates' competing tax plans.

The response given by Governor Bush (soon to be President Bush) veered and rambled all over the place. He never specifically addressed the question of how a 34-year-old single person would be affected by his tax plan . . . except at the very beginning of his ramble with a broad, general assertion: "You're going to get tax relief under my plan," which didn't explain how much relief or what kind.

Governor Bush began by talking about offering an increased tax exemption to those who care for "an elderly," forgetting or ignoring the fact that the person who asked the original question specified that she had *no dependents*.

Next, he skipped to Medicare (a subject of less-than-immediate interest to a 34-year-old). Then he skipped further off the path on to the topics of world peace, military policy, education, and, finally, work ethic.

Six subjects later, in his last sentence, probably recalling that the original question dealt with taxes, Governor Bush belatedly referred again to "tax relief in the proposals I just described," skimming over the fact that he never did describe any tax proposals.

This is not meant to pick on George W. Bush. Some of our most effective political leaders have been known to speak in a rambling fashion (Dwight D. Eisenhower, for one). And, speaking crisply and logically is no guarantee of statesmanship or political wisdom. Depending on your own political views and personal tastes, you might find Bush's speaking style infuriating, comic, or refreshingly human.

The essential point is a more general one. An excerpt of spoken language, when transcribed and printed, will never read like well-crafted prose. As a personal example, I recently recorded myself during a program with my clients, delivering the same material I've delivered for two decades. When I read the transcription, I was surprised to see how irregular my word patterns were. The reason: Spoken language is governed by the right brain. Rather than focusing on the rules of logic, grammar, syntax, and consistency, the mind flows freely, wherever the concepts lead.

By contrast, the production of written language tends to be governed by the left brain. When most people sit down to write a letter, memo, or report, their minds are front-loaded with those left-brain functions: logic, grammar, spelling, and punctuation. Rather than bouncing freely from one idea to the next, considering names, references, and concepts that may or may not be clearly developed, they move methodically through a sequence of points, meticulously self-correcting their syntax and logic as they go.

The result is a document that is technically correct. It doesn't contain fractured sentences like George W. Bush's "You'll be in a world in which—fits into my philosophy, you know, the harder work—the harder you work, the more you can keep," or repetitions like his "Now, I hope people understand that Medicare today is—is—is important."

If the writer's thinking is being ruled by the left brain, the natural flow of concepts is often impeded. As a result, the ensuing document almost inevitably omits ideas that are necessary or, worse, includes ideas that are unnecessary, overly detailed, or irrelevant.

You've probably written documents this way yourself: sitting down cold at your keyboard and banging out an email, memo, or letter, "editing" it for style and content on-the-fly. If so, you may have had the experience of reading the document afterward and discovering that you'd completely forgotten to mention an important fact or idea, or you may have stuck in a completely irrelevant detail. This is a natural by-product of left-brain dominance.

Developing a presentation by starting with left-brain considerations such as logic, sequence, grammar, and word choice (or, for that matter, the color, style, and design of slides) is ineffective. Crafting a presentation is a creative task, and it must start with the resources that are available only on the right side of your brain. Use the right tool for the right job.

Crafting a presentation is a creative task, and it must start with the resources that are available only on the right side of your brain.

Therefore, begin your story development process by doing what your brain is going to do anyway: follow the stream of consciousness, and capture the results during Brainstorming.

Managing the Brainstorm: The Framework Form

When you start to Brainstorm about your very important presentation to your very important client, do you want to start by thinking about what you'll wear? I doubt it. Do you want to start by thinking about the rest of your schedule on the day of your important presentation? I don't think so. Attire and calendar are related to the presentation, but only peripherally. You needn't include them in your Brainstorm.

Before you begin the Brainstorming process, you must first tighten your focus past the peripherals. To do this, begin with a tool called the *Framework Form*.

Think of your presentation as a blank canvas within a frame. This is where you will do your Brainstorming. To tighten the focus, you need to set the outer limits, the parameters, of your presentation. They include the following elements.

Point B

Since most presentations lack a clear point (the first of the Five Cardinal Sins), why not start with it? In other words, start with the objective in sight, and work toward it. Here again, this rule incorporates the wisdom of Aristotle and Stephen R. Covey.

Audience

Now that you appreciate the importance of Audience Advocacy, you must analyze what your intended audience knows and what they need to know to understand, believe, or act on what you are asking. Use the following three metrics to analyze your audience:

- **Identity.** Who will be in the audience? What are their roles?
- **Knowledge level.** Remember that one of the Five Cardinal Sins is being too technical. You cannot be an effective audience advocate unless you know your audience and are prepared to communicate with them in language they understand. Therefore, it's important to spend time during your preparation process analyzing your audience and anticipating what they know and what they don't know.

  As a tool for assessing your audience's knowledge level, I've developed a simple, nonscientific chart, called the comprehension graph. It measures knowledge along the vertical axis, from zero (no background knowledge about the presentation topic) to the maximum (knowledge that usually only the presenter would have). The horizontal axis measures the number of people in the audience.

To use this graph, mark points along it that represent what fraction of your audience will be located at each knowledge level along the vertical axis. For example, for a presentation about a new high-tech product to an audience that includes a large number of relatively unsophisticated listeners, along with a handful of engineers and other knowledgeable experts, the graph might look like Figure 3.1.

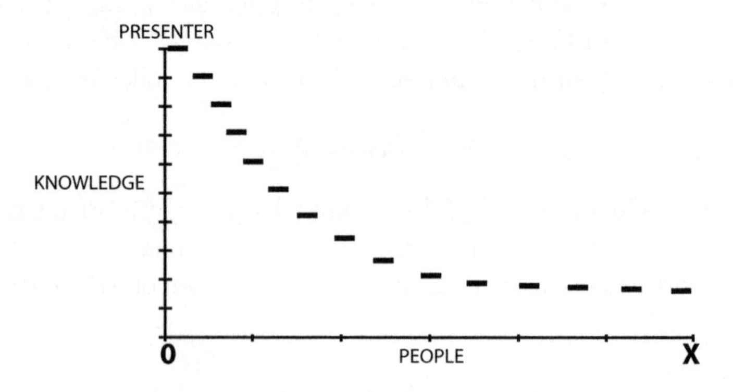

Figure 3.1 *The comprehension graph.*

The specific shape of the line you draw should be in your mind constantly as you prepare and present your material. If only a few members of your audience share your level of technical or industry knowledge (as is often the case), you can't fly too high for too long. You need to put significant effort into translating your technical information, using language, examples, and analogies that everyone can understand.

When technical terms are unavoidable, you can raise the audience's knowledge level through the use of *parenthetical expansion*. Simply stop your forward progress and explain your complex concepts and terminology by parenthetically adding, "By that I mean . . . " and then going on to offer a clear and simple definition.

- **The WIIFY.** This is undoubtedly the most important factor in analyzing your audience. Remember that another of the Five Cardinal Sins is no clear benefit to the audience. Ask yourself: What does your audience want? How does the subject of your presentation offer it to them? How can you make the benefits to your audience crystal clear?

External Factors

External factors are conditions that are "out there," in the world, independent of you and your audience. They can impact your message and how it may be received. Some external factors are positive, and some are negative. For instance, when making a pitch for investment dollars for your company, a rapidly expanding market for your product would be a positive external factor, while the emergence of powerful new competitors would be a negative external factor. You must consider all the external factors as you prepare your presentation. In some cases, you may need to change the content of your presentation or alter its structure to respond to unusually powerful factors. Be sure that you include counterpoints to the negative factors.

Setting

Throughout the preparation process, keep in mind the physical setting for your presentation. These factors, too, may affect the content of your presentation. You can analyze the setting by asking and then answering these classic journalistic questions:

- **Who?** Will you be the only presenter? If not, how many others will be presenting before and after you? How will you distribute the parts of the story among the presenters?
- **When?** When will you be making the presentation? How much time will you be allotted? Will you have time for audience interaction? Will there be a question-and-answer period?
- **Where?** Will you be meeting in your company's offices, on your audience's turf, or in some neutral setting? How will the room be arranged? Will it be an intimate or massive setting? In a larger setting, where only you will have a microphone, it's likely you'll give the entire presentation uninterrupted. In a smaller setting, interruptions are inevitable. If so, allow time for discussion.
- **What?** What kind of audiovisual aids will you be using? Will you be doing a demonstration to show your product in action? If so, will there be room and visibility to perform the demo? When will you do the demo: before, during, or after the presentation?

You must determine all these factors and define them in your Framework Form before you start your Brainstorming. Use the Framework Form shown in Figure 3.2 to assemble and capture all the preceding information as you develop it.

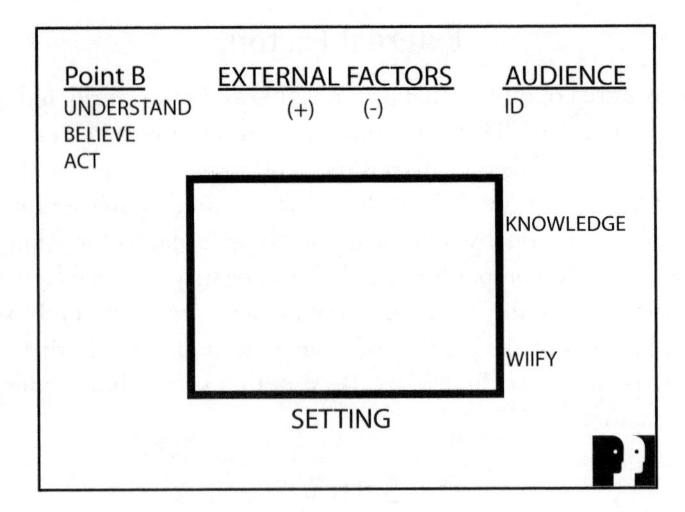

Figure 3.2 *The Power Presentations Framework Form. (You can download a copy of this form by visiting our website, www.powerltd.com.)*

Define all these factors as clearly and as specifically as possible. There is no such thing as a one-size-fits-all presentation. Build a presentation tailored to *one* audience, on *one* occasion, presented by *one* set of presenters, conveying *one* story, with *one* purpose. A presentation that is not custom-built will inevitably be less effective and less likely to persuade. Why bother presenting at all if you are not prepared to invest the time needed to make your presentation all that it can be?

Does this mean that you need to start every presentation from scratch? Not necessarily. After you've done the process once, the second time will be much shorter, and shorter still for each succeeding iteration. In the 20 years I've been in business, the process has never taken me more than an hour and a half the first time with a client, regardless of the subject, from the most complex biotech company to the simplest retail story. It usually takes 15 minutes the second time, and less each time thereafter.

Eventually, you'll be able to click and drag parts of one presentation into another. The key is to develop each presentation by starting with the basic concepts of the Framework Form. This initializing process will help ensure that each presentation you make is effectively focused on each persuasive situation you face.

Resist any temptation to skip or short-circuit the Framework Form process. Don't take it for granted that everyone in your group knows and understands the mundane who-what-when-where-why details of your presentation. Lay out these facts in advance; they will have a positive impact on what should and should not appear in your presentation, and in what form.

Now that you've set the context and focus, you're ready to begin developing potential concepts. Now your right brain, rather than thinking about attire or schedule, can focus on more relevant ideas. You'll want to capture those ideas as they emerge, and that's where we turn to Brainstorming.

Brainstorming: Doing the Data Dump Productively

Here's how to do productive Brainstorming:

1. Set up a large whiteboard or an easel with a big pad of paper and lots of pushpins to mount the sheets. I prefer a whiteboard because it allows me to erase and rewrite free-flowing ideas at will. It also results in a neater and easier-to-read set of Brainstorming notes. Have on hand a supply of markers in several colors. Use different colors to indicate different groups or levels of ideas.

   There are several high-tech products on the market that can capture written scrawls electronically from a whiteboard to a computer and then to a printer. (I use eBeam, www.e-beam.com.) These tools are very cool, but not essential. You can always ask someone to hand-copy the notes during or after the Brainstorming.
2. Gather your Brainstorming team. It should include all those who will participate in the presentation, as well as any others who have ideas or information to contribute.
3. You, as the presenter, or someone from your group (with reasonably neat handwriting) should handle the markers and capture the Brainstorming ideas on the whiteboard. This person is your scribe. In my programs with my clients, I act as both scribe and facilitator. As a facilitator, I assume a neutral point of view and simply take down all ideas as they come up, without judgment. There are no bad ideas in Brainstorming. Let them all flow. That is the essence of right-brain thinking. I also ask that each person in the group feed his or her ideas through me so as not to lose any ideas in side discussions, crosstalk, or digressions. I post all the ideas on the whiteboard for all to see and share.

There are no bad ideas in Brainstorming.

Have your scribe assume a similar role. Your scribe should not have a bias for or against any idea that emerges. Consider your scribe as Switzerland: neutral in all events.

4. Launch the Brainstorming session by having someone, anyone, call out an idea about something that might go into the presentation. One person might say "Management." You or your scribe should write the word "Management" on the whiteboard and then draw a circle around it to turn that concept into a self-contained nugget.
5. As each concept comes up, the entire group should help to *explode* the concept. For example, once "Management" appears on the whiteboard, pop out whatever ideas come to mind that are related to management. For example, there are the various members of your company's top management team: the CEO, the chairman, the CFO, the executive vice president. You or your scribe should jot these down as they come up, circle them, and link the circles to form a cluster of related ideas. Call the major idea in a cluster the "parent" and the subordinate ideas connected to it the "children."
6. Continue to do the same for other concepts that people in the group suggest. Certain concepts come up in almost every business presentation: "Our Products," "Our Customers," "Market Trends," and "The Competition." Depending on the specific purpose of your presentation and the issues your company is currently facing, some concepts will be unique to your presentation. As you work, you'll gradually fill the whiteboard with related concepts that might look something like Figure 3.3.

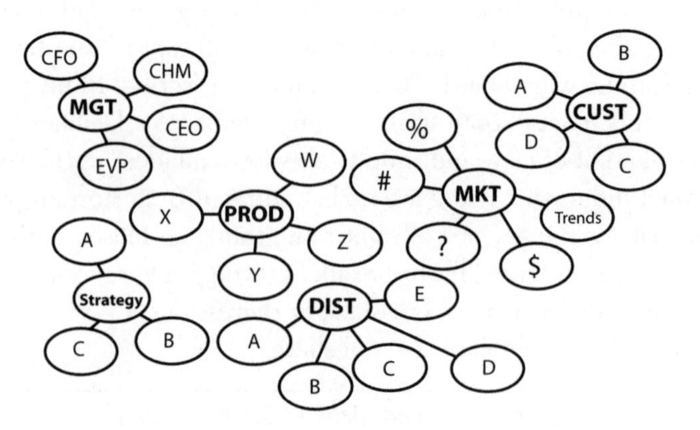

Figure 3.3 *A typical Brainstorming whiteboard.*

7. As you work, be flexible! Don't be afraid to bounce from concept to concept as necessary. While the group is exploding the concept of "Marketing Plan," someone might interject, "Oops! We forgot to list Jim, the marketing vice president, as a member of the management team." No problem; squeeze Jim in on the whiteboard. If necessary, use the eraser.

   Someone else might say, "There's a market statistic I'd like to include, but I'm not sure the latest data is available." No problem; note the idea wherever it belongs with a question mark in the circle. The placeholder will remind you that further research is needed.

As the Brainstorming proceeds, you'll find that ideas pop up all over the place. The ideas will shift, connect, disconnect, and duplicate as they seek relationships with other ideas. This is your right brain at work. As ideas continue to come up, they will move around. Let it happen. Relationships will emerge, change, and develop. Capture all the activity on the whiteboard.

The Spirit of the Brainstorm

While your team is Brainstorming, the right brain must rule. Remember that most businesspeople are left-brain-oriented, conditioned by education and experience to apply logic, reason, and rules to every activity. Learn to stifle this tendency during your Brainstorming. Avoid wordsmithing ideas. If you get bogged down in debating the proper words, you'll impede the free flow of fresh concepts. It's hard to avoid wordsmithing at first, but you'll find it surprisingly liberating.

Remember: There are *no* bad ideas in Brainstorming. Avoid censoring *any* ideas. The person whose idea is rejected is likely to feel rebuffed and may become reluctant to offer other ideas. When anyone mentions a new idea, jot it somewhere on the whiteboard, even if it strikes others as trivial or irrelevant. Even a seemingly needless idea can be useful, since it may stimulate someone else to bring up a related fact that may turn out to be important. Get it *all* down. Don't worry about recording too much information; not everything on the whiteboard will end up in your presentation. Consider all ideas during Brainstorming as candidates, not finalists. The *right* time to do the Data Dump is during your preparation . . . and not during your *presentation*!

Consider all ideas during Brainstorming as candidates, not finalists.

Avoid thinking about structure, sequence, or hierarchy. If you find yourself wanting to say, "That idea should go up front" or "That idea should close the presentation" while other ideas are popping up, it would be like trying to rub your stomach and pat your head at the same time. Structuring front-loads your mind with sequence, order, and linear thinking, the hallmarks of your left brain. Instead, let the concepts tumble out in nonlinear fashion, just the way the synapses of your brain fire naturally. Think about structure later. Remember: Focus *before* Flow.

Give yourself enough time to do a thorough Data Dump. Don't put down your markers the first time there's a long pause in the conversation. Chances are the group is just taking a mental breather. Most Brainstorming sessions feature two or three "false finishes," each followed by an explosion of new ideas, before the group has really exhausted its store of information and ideas.

When you are truly done, your whiteboard will be filled with lots of circles. At that point, the entire group will be able to see all the elements of your story, all the candidate ideas, laid out for easy examination and organization.

If any of this sounds familiar, it should: It is the kind of out-of-the-box thinking that many businesspeople use in strategic planning, product development, or problem-solving sessions. Well, these are the very same minds and the very same subject matter that go into a presentation. Why not use the same process?

One of the benefits of Brainstorming is that it provides a panoramic view. It's like spreading out all the parts of a kid's bicycle before you start trying to follow the all-too-complicated assembly directions; or the way a chef lays out all the ingredients for a complicated dish before the cooking begins in what's called a *mise en place*. Spreading out the raw materials of your presentation gives you ready access to and control of all your ideas.

Contrast this approach with a left-brain, linear process. The typical left-brain method is to start by designing Slide 1: "Okay, we'll open with our company mission statement"; then Slide 2: "Now let's talk about the management team"; then Slide 3: "Now the statistics about the marketplace"; and so on. The problem with this approach is that, as you focus on the slides one by one, each slide effectively covers and hides the slide before. As a result, you're looking at only one concept at a time. You never see the whole story at once; therefore, you never see all the components organized into a few key high-level units.

Instead, the Brainstorming approach follows the right brain's natural functions. It allows your ideas to pour out in a random, nonlinear fashion, ensuring that every relevant concept (as well as every irrelevant one) gets a place on the radar screen. Later, you'll enlist the help of the left brain in bringing order to the raw materials you've generated.

Roman Columns: The Technique of Clustering

Brainstorming generates a host of ideas of varying importance, loosely related to one another. The first step in getting from this relative chaos to an organized, clearly focused presentation is a technique known as *Clustering*.

Actually, we've already used Clustering to a degree. In the previous group Brainstorming example, every time the group exploded a concept into a series of related concepts, forming a group of linked circles on the whiteboard, they created a cluster. These clusters reflected the natural relationships among the ideas as they poured out during Brainstorming: *parents* and *children*.

Clustering is a necessary technique for organizing any complex material for presentation to an audience. It's also an ancient concept, dating back to the classic rhetoricians of Greece and Rome.

There's a story, probably apocryphal, about a Roman orator whose memory was legendary. (It may have been Cicero, although the documentation is sparse.) The orator often spoke in the Roman Forum extemporaneously for hours, without referring to a single note. His secret was a memory technique that is still used today. We can imagine him explaining it to a curious admirer in a dialogue like this: "You asked me how I can speak coherently at length, without written notes. Did you notice today how I walked around the Forum as I spoke?"

"Indeed I did. I assumed you did so in order to reach out to those in every corner of the audience."

"In part," replied the orator. "But there was a more important reason. As I walked from point to point around the edges of the Forum, I paused for a time at six different marble columns. Those columns are my memory aids. Each one symbolizes and reminds me of one group of ideas. Thus, rather than memorizing dozens of particular details, I have to recall only the six key ideas. Each of those key ideas evokes the details related to it."

Did Cicero really use this technique 2,000 years ago? No one knows for sure. But today I urge my clients to use the same technique to distill the ideas in their presentations. Clustering lets you reduce the 40 or 50 ideas that fill your whiteboard to five or six Roman Columns, the key ideas that organize all the rest. Each column has a group of subordinate ideas. Now instead of trying to organize many ideas at the detail level, you can organize them at the 35,000-foot level.

When you look at your whiteboard filled with ideas, you will find key clusters emerging from the chaos. Examine the whiteboard, and use a new marker color to highlight the most significant ideas. The idea is to make the parents stand out visually from the mass of data, as shown in Figure 3.4.

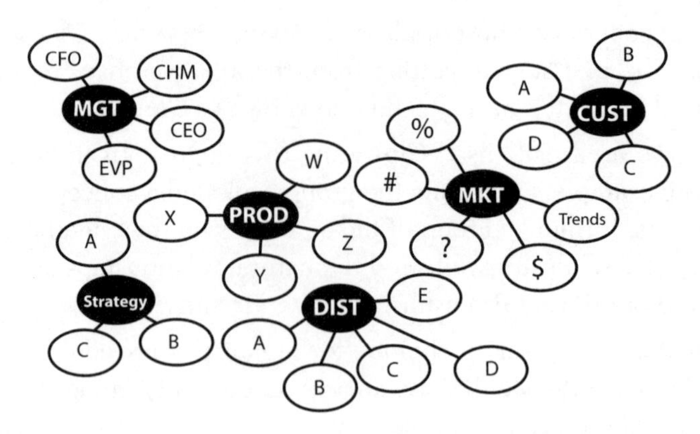

Figure 3.4 *Parents and children.*

As your group works on identifying clusters, you may find yourselves identifying links and connections that didn't occur to you before. That's fine; just draw lines on the board as needed, or erase and redraw the circles and lines as necessary. You may find yourself shifting concepts around: "Say, doesn't that point about the changing demographics of our market belong with 'Key Trends' rather than with 'Sales Potential'?" "How about connecting 'Cost Savings' to 'Customer Benefits' instead of to 'Unique Product Features'?" No problem; move the children and link them to the most appropriate parent.

If some ideas seem to have no connection to any of your Roman columns, now is the time to ask whether those ideas are truly relevant and necessary. Perhaps they don't deserve to survive the transition to the finished presentation. And if you think of new ideas that should be inserted, go ahead and add them. That's all part of the process.

As you can see, the technique of Clustering begins the process of organizing and introducing logic into the presentation. After having deliberately held back your left brain, you can now let it begin to get into the act.

Splat and Polish

You may be tempted to short-circuit the process by skipping the Brainstorming stage. "Why not *start* with clusters of key ideas?" you might say. "I could probably sit down right now and list the five main points we need to emphasize. That would save us all a lot of time." That's your logical left brain speaking. It wants to avoid the messy, uncontrolled process of free association. But the human mind doesn't work that way.

Start by unloading a "Splat!" of ideas in whatever order they came out, free-form . . . a total Data Dump. Organize them later, and later still polish them into words and sentences and paragraphs and, ultimately, into slides. This process is called *Splat and Polish*.

In my many years in the media, I've learned that most professional writers . . . novelists, journalists, playwrights, technical writers, and historians . . . follow this same process. Not one of them will write a single page of text until they've done their research, brooded over their topic, and assembled a mass of notes about it. They may note their ideas on Post-its, on dog-eared index cards, in spiral-bound notebooks, or simply in stacks of loose pages. Those notes, of course, are their Data Dump.

Results-oriented businesspeople, unfortunately, don't use the same process when creating a presentation or, for that matter, when writing a report, a speech, or a memo. That's the way businesspeople are accustomed to thinking: Get to the endpoint as quickly as possible. Find the shortest distance between two points. They figure that the quickest way to get a presentation done is to just start writing. Logical, yes? Yes, and wrong.

Here's a story that illustrates the pitfalls that the Splat and Polish approach can help you avoid:

Judy Tarabini (now McNulty) was a vice president in the technology unit of the Hill and Knowlton Public Relations Agency when Ben Rosen, continuing his promise to help me grow my business, introduced me to the firm. After I delivered my program successfully to one of Judy's clients, she began to call on me regularly for her other clients.

In 1993, Judy joined the corporate communications department of Adobe Systems. It wasn't long before she called on me to work with Adobe. This time she had a high-level, mission-critical presentation: Adobe was about to introduce its Acrobat product, and they were planning to have their entire senior management team, about 15 strong, fan out into the market to make launch presentations. Judy was so positive about my program, she convinced Adobe's entire senior management team, including the founding chairman and CEO, John Warnock, and his cofounding partner and president, Chuck Geschke, to participate in a story development session with me.

As always, we started with a blank slate. I stepped up to the immaculate whiteboard in the amply appointed executive conference room at Adobe's then-brand-new corporate headquarters in Mountain View. (They have since moved to even newer and more advanced facilities in San Jose.)

I started drawing out the executives. We began with Point B, we continued on to the WIIFY, then we moved on to the Brainstorming. As those very bright and very high-powered people spouted their thoughts, I raced to capture them on the whiteboard. We got lots of clusters: the Acrobat rollout schedule, the distribution plan, the Acrobat partners, the product benefits, the market, and many more. Before long, the whiteboard was filled to the edges with clusters of ideas.

Then there was a pause. I looked around the room and said, "Please take a moment and look at all the clusters on the whiteboard. Tell me whether we need to alter any of the ideas, whether we need to consider shifting associations, or whether we've omitted anything."

A thoughtful silence ensued. Then suddenly, reverberating in the silence, there was a sharp *thwack*! Chuck Geschke had slapped his palm against his forehead, as in the "I should've had a V8!" television commercials. Then he broke into a sheepish grin and said, "We've left out what Acrobat *does*!"

Does that sound odd? Sure it does. But it happens a lot. You're so close to your business that it's easy to take key ideas for granted, or to overlook or forget about concepts that are second nature to you but unfamiliar to your audience. That's one huge reason why you should never try to end-run the Brainstorming process and rush past story development. Take the time to make certain that everything (and that means *everything*) that may be relevant has had a chance to surface. Realizing what you omitted five minutes before the start of your presentation will be too late!

Focus *Before* Flow

Notice that so far I've held off any discussion about selecting the sequence of ideas for your presentation, which is the left-brain process. First, get the heart of your story straight; then and only then can you think about the most effective sequence of concepts for presenting that story persuasively. In developing your presentation, keep the proper order of the creative process in mind: Focus *before* Flow.

Now, having set the context with the *Framework Form*, having poured out all the concepts that might be relevant to your presentation by *Brainstorming* (an efficient Data Dump), and having condensed those concepts and ideas by *Clustering*, you're ready to develop the flow of ideas that will guide your presentation. That's the topic of the next chapter.

Chapter Four

Finding Your Flow

COMPANY EXAMPLES

- **Intel**
- **Cisco Systems**
- **BioSurface Technology**
- **Tanox**
- **Cyrix**
- **Compaq Computer**
- **ONI Systems**
- **Epimmune**

In the first three chapters, you worked your way through the first three steps in creating an effective presentation. You developed the Framework Form, including your Point B, and you determined your audience's WIIFY and their knowledge level. You brainstormed potential ideas to include in the presentation and distilled those ideas into clusters, also known as Roman Columns. You achieved right-brain focus.

Now you're ready to shift to your left brain and put your clusters into a sequence, to develop a logical flow. It's time to decide which Roman column goes first, which goes in the middle, and which goes last. In other words, you need a clear path that links all of your Roman columns, a blueprint for determining the best order for the elements of your presentation. You need *flow*.

The best way to express the critical importance of flow to your audience is to start with the simple example of written text. One distinctive aspect of written text is that the reader, who is the audience to the writer, has *random access* to the writer's content. If the reader, while browsing through a book, report, or magazine, encounters a word or reference that is unclear but looks familiar, the reader can simply place a finger in the current page and then riffle back through the prior pages to find the original definition or reference. The reader can *navigate* through the writer's ideas independently.

Your presentation audience does not have that capability. They have only *linear access* to your content, one slide at a time. It's like looking at a forest at the level of the trees, only one tree at a time.

You may be doing an excellent job of presenting one tree. Your audience may be suitably impressed, thinking, "That's a superb tree: deep roots, thick bark, rich foliage!" But if, when you move on to the next tree, you don't make it crystal clear how it relates to the first tree, your audience is forced to try to divine the relationship on their own. They no longer have access to the first tree, which forces them to work harder to remember it and draw the necessary connections.

At that point, your audience has three choices:

- They can fall prey to the MEGO syndrome.
- They can interrupt you to ask for an explanation.
- They can start thinking in an effort to understand the missing link, and stop listening.

None of these options is acceptable. *Don't make them think!*

Your job, therefore, is to become the navigator for your audience, to make the relationships among all the parts of your story clear for them. Make it easy for them to follow, to bring them up from the level of the trees and give them a view of the entire forest.

Doing this requires a road map, a plan, a formula. Like a chef who follows a recipe to use the right ingredients in the right order, you need to encompass all the Roman columns within an overarching template.

There are proven techniques for organizing ideas in a logical sequence to create a lucid and persuasive presentation. These techniques are called *Flow Structures*, and there are 16 different options for various types of presentations.

The 16 Flow Structures

1. **Modular.** A sequence of similar parts, units, or components in which the order of the units is interchangeable.
2. **Chronological.** Organizes clusters of ideas along a timeline, reflecting events in the order in which they occurred or might occur.
3. **Physical.** Organizes clusters of ideas according to their physical or geographic location.
4. **Spatial.** Organizes ideas conceptually, according to a physical metaphor or analogy, providing a spatial arrangement of your topics.
5. **Problem/Solution.** Organizes the presentation around a problem and the solution offered by you or your company.
6. **Issues/Actions.** Organizes the presentation around one or more issues and the actions you propose to address them.
7. **Opportunity/Leverage.** Organizes the presentation around a business opportunity and the leverage you or your company will implement to take advantage of it.
8. **Form/Function.** Organizes the presentation around a single central business concept, method, or technology, with multiple applications or functions emanating from that fundamental core.
9. **Features/Benefits.** Organizes the presentation around a series of your product or service features and the concrete benefits provided by those features.
10. **Case Study.** A narrative recounting of how you or your company solved a particular problem or met the needs of a particular client and, in the telling, covers all the aspects of your business and its environment.
11. **Argument/Fallacy.** Raises arguments against your own case and then rebuts them by pointing out the fallacies (or inaccuracies) that underlie them.
12. **Compare/Contrast.** Organizes the presentation around a series of comparisons that illustrate the differences between your company and other companies.
13. **Matrix.** Uses a two-by-two or larger diagram to organize a complex set of concepts into an easy-to-digest, easy-to-follow, and easy-to-remember form.

14. **Parallel Tracks.** Drills down into a series of related ideas, with an identical set of subsets for each idea.
15. **Rhetorical Questions.** Poses, and then answers, questions that are likely to be foremost in the minds of your audience.
16. **Numerical.** Enumerates a series of loosely connected ideas, facts, or arguments.

You can use these Flow Structures to group your clusters in a logical progression, making it easy for your audience to follow your presentation and easy for you to construct your presentation. To do that, let's look more closely at each of the 16 Flow Structures, with brief examples of each. The chances are excellent that one or more of these Flow Structures can serve you well in organizing virtually any presentation.

1. Modular

A presentation organized in modular form is a sequence of similar parts, units, or components in which the order of the units is interchangeable. Think of the Modular Flow Structure as a plug-and-play approach: The presenter makes an arbitrary decision about the sequence of units and then presents them to the audience, one by one. This is the most loosely organized of the 16 Flow Structures, which can make it challenging for your audience to follow.

Sometimes a presenter has no other choice than to use the Modular option. Financial presentations fall into this category . . . even a financial module within a larger presentation. In an IPO road show, the CEO usually uses one of the other, more accessible Flow Structures to organize the main components of the presentation in a meaningful order. But the CFO, who presents the annual results, quarterly results, balance sheet, income statement, and other financial data, can do them in almost any sequence. It hardly matters whether the annual results precede or follow the quarterly results. The CFO can simply order the items in whatever sequence feels right and discuss each one in order.

There are other instances in which the Modular option is appropriate. For example, it can be employed for a product introduction presentation that consists mainly of new product features.

The Modular Flow Structure provides a couple of advantages. If need be, you can easily rearrange the items at will. Or, if you're faced with time constraints, you can even omit one or two items. Convenient, yes, but challenging for your audience to follow and for you to deliver. Because there's no compelling *logic* to the clusters, everyone (that includes you *and* your audience) is under great pressure to try to track them. Therefore, use the Modular option sparingly and briefly; and if you do, add continuity with the Linkages you'll find in Chapter 10, "Bringing Your Story to Life."

2. Chronological

Far more accessible than the Modular option is the Chronological Flow Structure, which organizes your clusters of ideas along a timeline. This option, reflecting events in the order in which they occurred or might occur, is ideally suited for any presentation where *telling a story that deals with change* is the most important objective.

You might have to make a presentation whose main purpose is to explain to the audience how a particular state of affairs came to be. For example, suppose you are the human resources director of a large company (let's call it "Goliath Software, Inc.") that has just purchased a smaller competitor ("David Software Co."). The announcement of the acquisition came as a surprise to nearly all the employees of both companies, and now many of them have questions and concerns: Why has this happened? How are the employees of these two companies, long-time rivals, supposed to begin working together? What do they have in common? What does the future hold?

Senior management has called a meeting to be held in the auditorium of David Software, the acquired firm, and you've been asked to make a presentation to "bring the new team up to speed." Point B, of course, is to help the David Software employees understand the acquisition and begin to feel as though they are a part of the new, expanded Goliath family of companies.

One effective way to organize your presentation could be the Chronological Flow Structure. Present a timeline showing where Goliath Software started, where it is today, how and why the acquisition occurred, and how the two companies will move forward together in the future. By making your audience conversant with Goliath's history and by showing them that the acquisition is a natural outgrowth of the convergence between the two firms, you are likely to assuage the concerns of the David Software team and to begin building strong connections with their new employer.

3. Physical

If the Chronological Flow Structure organizes your presentation according to the logic of time, the Physical option follows the logic of place. A presentation using a Physical Flow Structure takes its cues from geography, organizing clusters of ideas according to their physical location.

Suppose your company is a distribution operation whose points of presence around the world represent its major competitive advantage. And suppose you are asked to make a presentation to an audience of potential customers who operate international businesses and are looking for a distribution partner that can serve their needs globally. You might organize your presentation according to the Physical Flow Structure.

You might introduce your clusters by saying, "Worldwide Distribution, Inc. has warehouses and shipping centers at 11 strategic locations on five continents, from the U.S. to Australia, from Brazil to France, from China to North Africa. To help you see why Worldwide is better positioned to serve you and your customers than any other distribution company, let me walk you through how each of those centers operates and connects within our global distribution network." This is an intuitive, easy-to-follow structure for the presentation.

4. Spatial

In contrast to the Physical option, which follows a literal geographic arrangement, the Spatial Flow Structure organizes your ideas conceptually, according to a physical metaphor or analogy, providing a spatial view of your topics: for example, from the top down, from the bottom up, from the center out, or from the outside in.

I often give presentations at industry or financial conferences, where I'm asked to discuss what it takes to create an effective presentation. For this purpose, I use a Spatial Flow Structure: from the bottom up, depicting this graphically as a pyramid, as shown in Figure 4.1.

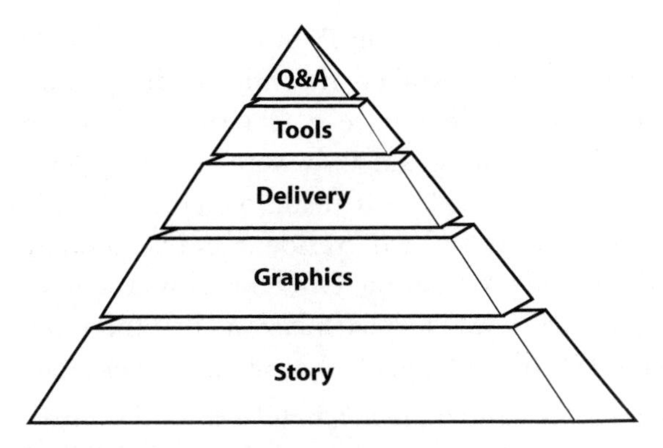

Figure 4.1 *Spatial Flow Structure: from the bottom up.*

The forest view is that every presentation has five primary components:

- Story development
- Graphics design
- Delivery skills
- Tools of the presentation trade
- Question-and-answer techniques

Point B of this presentation is that a really effective presenter masters all five components. Of these five, the most fundamental is the story. Starting with the story at the foundation, I then go on to describe all the other components, working my way up the pyramid to the peak, the question-and-answer skills. Along the way, I provide techniques to optimize each component, illustrating with examples and anecdotes.

The techniques and illustrations vary from one conference to another, but the Spatial Flow Structure remains the same. It creates an image that audiences find easy to understand, remember, and transmit to others: "What did Jerry say about the effective presentation?" "He said it's like a pyramid, with the story at the bottom and question-and-answer skills at the top."

The Spatial Flow Structure can be used the other way as well: from the top down. Here's an example:

At Intel Corporation, Randy Steck and Dr. Robert Colwell were the leaders of a crack engineering team working in an atmosphere of the utmost secrecy to develop Intel's next-generation integrated circuit, the P6. (The *San Francisco*

Chronicle wrote, " . . . you'd think the P6 was a tactical bomber rather than a computer chip." The episode recalled the title of the popular business book authored by Andy Grove, Intel's former Chairman of the Board: *Only the Paranoid Survive.*) When Randy and Bob were ready to unveil the chip, they brought me in to help them and their team, including then-General Manager Dadi Perlmutter (now Executive Vice President and General Manager of Intel's Mobility Group) and then-Marketing Director Lew Paceley, to develop the launch presentation, which was to take place at the International Solid State Circuit Conference (ISSCC), a highly technical enclave of their peers.

The design team sequestered me in a hotel a short distance away from Intel's ultra-high-tech facility in Hillsboro, Oregon, where I led Bob, Randy, Lew, and Dadi through the steps you read about in the previous chapters, including the Framework Form, Brainstorming, and Clustering, until we distilled all their ideas into the following clusters, or Roman Columns:

- Design Rationale for the P6
- P6 Product Specifications
- Potential End-User Products
- System Architecture and Supporting Chips

The Design Rationale described the technology at its highest level (the concept behind the design). The Product Specifications described the result of the design. The Potential End-User Products described the applications of the P6. The System Architecture described how all the components worked together. Each layer drilled down a little deeper, from the top down, creating a Spatial Flow Structure.

After the presentation at the ISSCC, Bob Colwell wrote to me with a post-mortem:

> *I really felt sorry for the other presenters, who were incredibly nervous, and did exactly as you predicted: classic data dumps on the audience . . . Because I knew the presentation cold, I wasn't at all nervous when I actually took the stage, despite having the audience so big (~1800 people) . . . The Q&A period was interesting. I didn't get any of the hardball questions I was worried about . . . I wonder if maybe potential hostile questioners were holding back, for fear of "losing" the argument to somebody who appeared to have done his homework and seemed thoroughly in control of the proceedings.*

The Spatial option has other uses. For example, the market for a product can be presented in terms of concentric circles, like those on a target. The heart of the market, is in the center circle (the bull's-eye) and represents those customers who are most certain to be interested in the product. The outer circles represent other markets that are larger, more diffuse, and increasingly hard to reach. The center-out Spatial Flow Structure organizes these ideas in a way that's easy for an audience to grasp, follow, and recall.

The simplest variant of the Spatial Flow Structure is the physical metaphor of constructing a house. The foundation describes the platform product or service. The supporting beams of the superstructure represent the organizations and partners. The wires and pipes of the internal infrastructure depict the technology. The glass, brick, and mortar of the external interface stand for the marketing and branding. Many different companies in many different industries have tracked their entire business model with this recognizable format.

5. Problem/Solution

The Problem/Solution Flow Structure is attractive because it has a built-in WIIFY. When you use this option, you organize your presentation around a problem and the solution offered by you or your company. In this option, the benefit your company has to offer through its product or service follows naturally.

Many companies in the life sciences (pharmaceuticals, genetic research, medical devices, health care) use this Flow Structure when doing a road show to raise private or public capital. To attract investors, they describe a particular medical problem and how they can solve it with their unique product or service. In their field, Problem/Solution is synonymous with illness/cure.

In the field of education, Problem/Solution is synonymous with learning. The essence of learning is to replace a lack of knowledge with skills. As a coach, I am also an educator, and so I use Problem/Solution in my programs, and in this book. Think back to the very first chapter. It began with "The Problem with Presentations." Everything since then has been my set of solutions for you. In fact, the entire Power Presentations program uses the Problem/Solution approach to address each component in Figure 4.1.

If you consider using Problem/Solution for your presentation, be careful about getting the emphasis right. Many people in business spend too much time on the problem and not enough time on the solution, leaving their audiences feeling as if they have slogged through a tragic Russian novel.

This matter of time and weight can best be illustrated with an analogy. In a Western film, the problem part (the Indian attack, the bad guys in the black hats, the tornado, the wildfire) lasts for a long time, virtually the full running length of the movie. That's because these problems produce suspense for the audience and empathy for the endangered protagonists. Those emotions are what keep the audience glued to their seats, and buying lots of popcorn. The solution, by comparison, takes a very short time: The U.S. Cavalry arrives and rescues the hero and heroine. The End.

However, if you decide to use Problem/Solution for your business presentation, shift the weight. Touch on the problem very briefly, then bring in the Cavalry troops of your solution and let them parade majestically, in splendid full-dress regalia, complete with a rousing marching band. After all, that's the part of your presentation you want your audience to remember most.

6. Issues/Actions

While virulent cancer and invasive surgery . . . along with man-eating sharks . . . are clearly problems, remember this caveat about using the Problem/Solution option in areas of business other than life sciences: Businesspeople don't like to be reminded of their problems. Every business today, from the smallest home office to the largest global enterprise, is under security attacks from hackers, spammers, viruses, and identity theft. These organizations are painfully aware of the negative impact on their costs, time, productivity, and confidentiality; they don't need salt rubbed in their wounds. So rather than remind your audience about their problems, describe their issues and tell them what actions you and your company propose to address them. This is more than euphemism; it shifts from a negative focus to one that is concerned and active. Audience Advocacy in action.

The Issues/Actions Flow Structure is frequently used for presentations by companies that are in a turnaround mode. They identify the issues they are facing and the actions they are planning to take to overcome them, producing a list of clusters that might look like this:

Outline: Turnaround Plan for Stumble & Falter, Inc.

Issue:	Out-of-Control Expenses
Action:	Immediate Cost-Cutting Measures and Hiring Freeze
Issue:	Unprofitable Product Lines
Action:	Sale of Three Business Units
Issue:	Flat Revenue Growth
Action:	Accelerated Development of Two Promising New Products

7. Opportunity/Leverage

A close cousin of the Problem/Solution and Issues/Actions Flow Structures is the Opportunity/Leverage Flow Structure. With this option, you begin by describing an attractive business opportunity: a huge new market, a change in technology, an economic shift, or some other driving force. Then in the leverage section, you describe the superior products, distribution methods, partnerships, or competitive strategy your company has developed to take advantage of that opportunity. Again, this is more than euphemism or mere semantic difference. This structure directs the focus to your audience's interests and how you can meet them. It is the embodiment of Audience Advocacy.

The Opportunity/Leverage option is the Flow Structure of choice for most IPO road shows because it appeals to the investor audience's essential interest in growth. You'll recall that I helped Cisco develop their road show presentation when they went public. At the time, Cisco's technology was an esoteric novelty. (Even today, as the established leader in computer networking equipment, Cisco's complex technology is still difficult for laypeople to understand.) The potential investor audiences for Cisco's initial offering did not yet fully understand how computer networks operated or why they would be so important in the future. Therefore, the Cisco road show team decided to start their presentation by demonstrating the enormous potential of networking *before* trying to explain the technology that did the networking. Thus, they chose the Opportunity/Leverage Flow Structure.

In their outline, the Cisco team began by describing the shift in computing from mainframes to PCs. This shift was neither a problem nor an issue; it was purely an opportunity. They then moved on to delineate the rapid growth of local-area networks and wide-area networks (LANs and WANs) and the recent improvements in technology that brought significant increases in speed, bandwidth, and power. They ended this cluster with a look at the anticipated shift in business from enterprise-centered to remote-based computing. All of those trends, taken together, represented the opportunity.

Next, they talked about how their new device, called a router, could internetwork all networks. They explained how Cisco manufactured the router, how they serviced it, how they sold it through channels and strategic relationships, and where they intended to go with the router in the future. All of these facts, taken together, represented Cisco's leverage of the opportunity.

Notice how this Flow Structure simplified and organized the presentation. Instead of having a dozen items for the presenter to explain and the audience to track, there were just two: Opportunity and Leverage. The forest view.

8. Form/Function

The preceding three Flow Structures (Problem/Solution, Issues/Actions, and Opportunity/Leverage) are close cousins. The Form/Function Flow Structure is distinctly different. It moves your company's business offering (its solution, action, or leverage) into the starring role, front and center. Use it when you're presenting a single central business concept, method, or technology that has many applications or functions emanating from that central core. Think: one core technology and multiple applications; a main theme and several variations; a hub and its radiating spokes; a foundation idea and its dissemination by way of multiple franchises.

A salesperson might use the Form/Function option when presenting any product or service that has multiple applications. For instance, the first salespeople who brought 3M's Post-it Notes to market might have used this Flow Structure to introduce the novel lightly sticking glue (the Form) and then gone on to describe its myriad uses (the Functions).

The Form/Function approach is often used by biotech companies because it not only brings the franchise science to the forefront, it organizes complex subject matter efficiently.

As an example, BioSurface Technology went public on the strength of a novel tissue engineering technology they had developed. (BioSurface was later purchased by Genzyme.) The BioSurface approach was based on the fact that the human body tends to accept autologous (self) cells or tissue as grafts because it recognizes such tissue as its own. Conversely, the human body tends to reject allogeneic (non-self) cells or tissue because it recognizes such tissue as foreign and therefore rejects the graft. Patients with major burns don't have enough of their own skin to be used as grafts. BioSurface discovered a way to take a postage-stamp-sized piece of a patient's skin and, in three weeks, grow it into enough autologous skin to cover the entire body surface.

In the BioSurface road show, CEO Dave Castaldi began by describing his company's innovative core science: how they extracted a patient's own cells, preserved them, cultured them, grew them, and then transplanted them back into a patient's own body, without rejection. This core tissue engineering technology represented their Form.

To demonstrate how it functioned, Dave then described how BioSurface was able to apply this science to a patient's own skin for permanent skin replacement, then to allogeneic skin from an unrelated donor for acceleration of wound healing, then to cartilaginous tissue, and finally to ocular tissue. One Form, multiple Functions. Of course, each Function represented a potential source of revenue and profits for BioSurface, and a business opportunity for potential investors.

From the sublime to the ridiculous:

Imagine that you're the CEO of Mom's Barbecued Chicken, seeking investment money to expand your business. In this case, Mom's barbecue recipe is the secret sauce, or the Form. The Functions would include all the ways the secret sauce could be developed as a business: by rolling out 600 franchised outlets, operating them with economies of scale, provisioning them with just-in-time deliveries, promoting them with co-op ads, and then by selling the secret sauce in 16-ounce jars in supermarkets and in plastic single-serving containers to airlines.

From the ridiculous *back* to the sublime:

Make it once; sell it many times; a classic business model represented by Gillette's razors and blades, Kodak's cameras and film, Hewlett-Packard's photocopiers and replacement toner cartridges. The cost is in the development of the core product; the profit is in the disposables: a high-profit-margin business.

9. Features/Benefits

This is the traditional product launch approach. In a presentation organized according to the Features/Benefits Flow Structure, you would discuss a series of features of your product or service, and for each one you would explain the concrete benefits provided to your customer. Once again, notice that the WIIFY is strongly woven into the very fabric of the presentation.

In the book business, each season's list of new book titles is presented by the publisher's sales representatives to buyers from the bookstore chains, like Barnes & Noble, as well as to buyers from individual, independent bookstores. For each title, the representative is expected to explain the book's specific features and the benefits it will provide to readers. A new atlas, for example, might boast features like larger type and brighter colors on its maps; the benefits to the readers are that the maps will be easier to read and use. The latest book in a series of thrillers might have as a feature "the most deadly and sinister conspiracy ever faced by Detective Cliveden"; the benefit is that the new book is a real page-turner. Fans of the series will spend several sleepless nights in delightful agony reading it.

If the sales rep presents book's features and benefits convincingly, the WIIFY and Point B will follow naturally. The rep can say to the bookstore buyer, "As you can see, these features and benefits are sure to make this new book one that dozens of your customers will be eager to buy" (the WIIFY). "That means you'll want to buy a lot of copies to stack in the front window of your store!" (Point B).

10. Case Study

A case study is essentially a story, a narrative recounting of how you or your company solved a particular problem, or how you or your company met the needs of a particular customer. In the telling, the case study covers all the aspects of your business and its environment. The Case Study Flow Structure provides a central spine that connects multiple diverse components.

We humans find stories, especially stories about people with whom we can identify, inherently interesting. Thus, a case study is an excellent way of capturing and keeping an audience's attention. It's an easy and practical way to make a product or service that is technically complex or apparently uninteresting become more vivid, personal, and understandable.

The human-interest angle is particularly applicable in medical business presentations. Let's say your Case Study is about a patient named John Smith. You can describe the illness John has contracted, how many other John Smiths there are in the world, how much money is spent on all those John Smiths, and how long they've suffered without a cure. Then you can talk about how your company's drug cured John Smith, the patents you have on the drug, its regulatory status, its clinical status, the cost of manufacturing it, its average selling price, and its potential profit margin. Finally, you can describe how John Smith was rehabilitated and reimbursed, thus explaining how your drug will sell in the managed-care environment. The story of John Smith provides a way to organize and humanize all the details of your company's entire story.

One of my IPO clients was a company that digitizes television commercials and then transmits and retransmits them from the advertising agency to the broadcaster and from the broadcaster back to the agency for audit. For the road show, we took one Dodge automobile commercial and followed it through the whole process, demonstrating all of the company's services and products as a superior alternative to conventional shipping methods. The Dodge case study served as the spine for the entire presentation, making the company's capabilities . . . and its potential as a business . . . tangible and convincing.

11. Argument/Fallacy

There may be times when you must make a presentation in the face of a highly skeptical or even downright hostile audience. At such times, consider using the Argument/Fallacy Flow Structure, in which you raise arguments against your own case and then rebut them on the spot by pointing out the fallacies (or inaccuracies) that underlie them. The idea is to preempt any objections in the minds of your audience, thereby creating a level playing field for a positive presentation of your company's real strengths.

This is a risky Flow Structure to use. It tends to sound either defensive or contentious, and it sets a negative tone. Reserve the use of this option for situations in which the negative ideas about you and your company are widespread and therefore unavoidable.

One company used this Flow Structure to its distinct advantage. The chairman of the board was scheduled to appear at a major investment conference to represent his company. He decided to title his presentation "Seven Reasons Why NOT to Invest in Us." He drew the seven reasons from negative analysts' reports, and one by one he rebutted them. When he was finished, the inescapable conclusion was that his company's stock was indeed a good buy.

12. Compare/Contrast

The point of the Compare/Contrast Flow Structure is to compare or contrast you or your company with others. How is your offering unlike that of any other company in your sector? How do you stack up against the competition? What is your competitive advantage? A presentation built according to this Flow Structure might focus on a series of comparisons, showing exactly what makes your company special along each parameter.

Like the Argument/Fallacy Flow Structure, choose this option with caution. By bringing another company into even partial focus, you run the risk of sounding defensive or, worse yet, having your audience remember the *other* company rather than your own. Moreover, when you attempt to throw a positive light on your own company by casting a negative light on another company, you may inadvertently offend someone in your audience who may have a direct connection with, or own shares in, the company you are criticizing.

For these reasons, save the Argument/Fallacy and Compare/Contrast Flow Structures for customer and industry presentations where you know your audience well, and there's less of a chance that you'll generate negative feelings that don't already exist.

13. Matrix

You're familiar with matrices: those two-by-two, three-by-three, or four-by-four boxes that can be used to organize items according to combinations of ideas or qualities. Business audiences love matrices. Perhaps it's because a matrix imparts a quasi-scientific feeling, or maybe it's because, like many of the other Flow Structures, a Matrix Flow Structure organizes a complex set of concepts into an easy-to-understand, easy-to-follow, and easy-to-remember form.

The Matrix is a close cousin of the Spatial Flow Structure in that it organizes concepts in a visual format. The difference between the two is that the Spatial Structure implies dynamic relationships or movement (top-down, bottom-up), while the Matrix implies stationary or stable relationships.

Figure 4.2 is an example. It shows how the market for personal financial services might be divided into four categories, based on divisions along two dimensions.

The two-by-two box creates a form you and the audience can follow through the whole presentation. You can analyze each of the four sectors in some detail, explaining why your company has chosen to focus on Sector 2 as the most promising sector in developing its business.

		NEED FOR FINANCIAL GUIDANCE	
		LOW	HIGH
INCOME	HIGH	Sector 1	Sector 2
	LOW	Sector 3	Sector 4

Figure 4.2 *A two-by-two matrix.*

14. Parallel Tracks

The Parallel Tracks Flow Structure is a compound form of the Matrix option. It drills down into each sector of a Matrix with identical subsets of information, or it drills down into each of a series of related ideas with an identical number of subordinate ideas.

Let me give you an example from the world of biotechnology, one of the most technically difficult of all subjects. Presenters need to work extra hard to find ways to simplify and organize their information. The Parallel Tracks option is an effective way of doing so.

Tanox, Inc., before its acquisition by Genentech in 2007, was a public biotechnology company that developed proprietary drugs to treat diseases that affect the human immune system, including asthma, allergies, AIDS, and others. For allergic diseases, Tanox's initial focus was to develop products that treat asthma, seasonal allergic rhinitis (ragweed and pollen), and severe peanut allergy. The founder and CEO, Dr. Nancy Chang, a Ph.D. in biological chemistry, often had to present complicated scientific processes to an audience of investors, giving them sufficient detail to convey the business potential of Tanox's science.

Nancy used the Parallel Tracks Flow Structure to arrange and clarify a complex story. First, she talked about how, in allergic patients, their bodies produce specific Immunoglobulin E, known as IgE, to the specific allergens that the patients are allergic to, and how IgE can trigger histamine release and cause all the disease symptoms of asthma and allergy. In Tanox's world, this is called the *allergic disease mechanism*. Then Nancy described Xolair, Tanox's *proprietary drug* that treats allergic diseases. She described how Xolair can potentially remove IgE from patients' bodies, thereby preventing the initiation of allergic disease and the resulting symptoms. In Tanox's world, this is called the drug's *mechanism of action*. Finally, she talked about the number of allergic patients who could potentially be treated by Xolair . . . in other words, *the market*.

Next, Nancy moved on to severe peanut allergy, AIDS, and others. One by one, she went through each disease with the same points: *the disease mechanism*, the Tanox *drug product* and its *mechanism of action*, and the size of the *market*. By the time she finished describing Tanox's product pipeline, her audience could almost sing along: *disease mechanism*, *product*, *mechanism of action*, and *market*. Thus, the Parallel Tracks Flow Structure makes a set of complex and technical facts easy for a lay audience to digest.

15. Rhetorical Questions

This Flow Structure may be the ultimate form of Audience Advocacy. It takes the audience's point of view, immediately involving them by saying, "You might be wondering . . . " and then answers the question for them. Of course, it's best if the questions are ones that your audience members are likely to have in their minds rather than ones that you strain to devise. The use of the Rhetorical Questions option is not effective if the questions are forced.

An entire presentation could be built around the Rhetorical Questions Flow Structure. Each of your idea clusters could be linked to a particular rhetorical question, to which you would then provide the answer.

Here's an example:

Cyrix, a company that designed, developed, and marketed semiconductors, went public in 1993. (Cyrix was later acquired by National Semiconductor and then by Via Technologies.) For the IPO road show, Cyrix CEO and co-founder Jerry Rogers decided to use the Rhetorical Questions Flow Structure.

In Jerry's opening statement, he said: "Cyrix competes against the established giants Intel and AMD (Advanced Micro Devices), as well as two other large, well-funded companies. So as a tiny start-up trying to establish itself, I have to respond to some challenging questions from potential new customers about the IBM-compatible microprocessors that Cyrix designs. Three questions keep recurring: 'Will Cyrix microprocessors run all software applications?' 'How will Cyrix compete with Intel?' and 'Does Cyrix have the financial stability to succeed?'"

Jerry continued, "Cyrix has already begun shipments of the first commercially available 486 microprocessors *not* produced by Intel. But those same three questions about compatibility, competition, and finances remain important to potential investors like you. In my presentation today, I'll provide you the answers to those questions to demonstrate that Cyrix is indeed a sound investment."

Jerry showed the three questions on a slide. Then he spent a few moments and a few slides providing an answer to each question. His final slide showed three declarative sentences, answers to the three questions.

In the course of answering the three questions, Jerry covered diverse topics such as chip architecture, manufacturing strategy, average selling prices, intellectual property rights, the prospect of litigation, and the competitive landscape. But he tied together all these topics into three Roman Columns, and then he tied together the Roman Columns in the Rhetorical Questions Flow Structure.

16. Numerical

Finally, there's that old standby, the Numerical Flow Structure. "There are five reasons why our company represents an attractive investment opportunity." Then you list the five, counting down as you go. Like the Modular option, this is a very simple, rather loose Flow Structure. Use the numerical format only if none of the other options works for your presentation.

Back in 1983, Compaq Computer Corporation chose the numerical option for their IPO presentation. At the start of their two-week road show, the team offered 10 reasons why an investor might want to buy shares of Compaq. But they soon found that all those reasons tended to be a drag on their investor audiences, with their occupationally short attention spans. The team quickly folded their 10 reasons into five and continued their road show to more attentive audiences.

The popularity of David Letterman's nightly "Top 10 List" brought the higher number back into favor. It also brought better luck for Hugh Martin, the then-CEO of ONI Systems, a provider of optical telecommunications equipment (now owned by Ciena Corporation), when he presented at the Robertson, Stephens, and Company Technology Investment Conference.

I attended the ONI presentation and took a seat at the back of the room, as I always do at such conferences, so that I can watch the audience as well as the presenter. Most of the time, because few presenters provide flow (one of the Five Cardinal Sins), the MEGO syndrome quickly sets in. The backs of the audience's heads start to shift about. First they glance down at their mobile devices, then they lean over to whisper to their neighbor, then look back to a copy of *The Wall*

Street Journal, and often, before long, they jump up and leave the room to attend a different presentation.

Hugh Martin avoided that fate. He began his presentation by saying, "I'm a fan of David Letterman, and so, in the same spirit, I'm going to structure my presentation along the lines of the 'Top 10 Questions' I get from institutional investors." As Hugh stepped through the 10 questions, and his responses to each of them, none of those occupationally hyperactive heads shifted, and no one left the room.

After the conference, Hugh gave me the following postmortem:

> *Robertson got enough positive feedback on that presentation that they are considering implementing the "Top 10" as a format. It really works, because every investor wants to know what every other investor is asking. It also works for me, the presenter, because I don't have to worry about flow. There is a natural break between each "Top 10" question, which means I don't have to tee up the next slide in advance. Once I've introduced the question, it's easy to keep flow through the answer. Overall continuity is there because everyone knows you're simply counting down from 10 to 1. It's a wonderful format because it works for the audience, and, truth be told, I can look really polished with little practice.*

Which Flow Structure Should You Choose?

Is one Flow Structure better than another? Not really. Some structures work especially well in particular situations. Opportunity/Leverage generally works well for investment presentations, and Form/Function generally works well for industry peer groups. But is there a right or wrong way to organize any particular presentation? No. In fact, the 16 Flow Structures overlap to some degree. The key is: Choose one or two Flow Structures for the *entire* presentation. Where presenters get into trouble is when they choose all 16 . . . and sometimes all 16 on each slide. Think of the 16 as choices on a restaurant menu. You select only one or two dishes from the menu. Select only one or two Flow Structures for your presentation.

Choose one or two Flow Structures for the entire *presentation.*

When you do make your choice, you'll elevate your audience from the trees to the forest. They'll be able to follow your presentation easily, and by navigating them through the parts, you'll be sending them the subliminal message of *Effective Management*.

If you fail to choose, your presentation will drift, your audience will become confused and drift too, and you'll never get them to Point B. It's less important *which* Flow Structure you choose than that you make a choice.

It's less important which Flow Structure you choose than that you make a choice.

Here's an example from the world of politics of how critically important it is to choose and follow a Flow Structure. It's the outline of the State of the Union Address[1] delivered by President Bill Clinton in 1995 after he lost his Congressional majority in the midterm elections:

1995 State of the Union Address

Change

Status

New Covenant

Accomplishments

New Proposals

Balanced Budget

Welfare Reform

Crime

National Service

Illegal Aliens

Tax Cut

Minimum Wage

Health Care

Foreign Relations

Responsibility

1 http://www.cnn.com/ALLPOLITICS/1996/resources/sotu/full.texts/1995.html

Can you tell which Flow Structure President Clinton was using? Don't worry; no one else could, either. There was none. In trying to be all things to all people, Clinton's choppy address prompted David S. Broder of the *Washington Post*, one of the nation's senior political commentators, to write, "It was a speech about everything, and therefore about nothing."[2]

That is not exactly the kind of review you want to receive for one of your presentations. Broder and most others in the audience suffered from full-blown MEGO syndrome. To make matters worse, at 82 minutes, the speech set a record as the longest State of the Union Address.

Apparently, President Clinton (or his speechwriting team) learned a lesson. The next year, this was the outline for his speech:[3]

1996 State of the Union Address

Strengthen Families

Education Goals

Economic Security

Fight Crime

Clean Environment

Lead Peace

Challenge Washington

President Clinton titled this speech "Seven Challenges," so the Flow Structure was that old warhorse, Numerical. Like all State of the Union Addresses, this was a complicated and lengthy speech, but the simplicity of the Flow Structure made it easy to follow. The audience . . . in this case, the nation . . . always knew where they were and where they were going.

Now let's flash forward to 2007 and look at the outline of the State of the Union Address[4] delivered by President George W. Bush after he lost his Congressional majority in the midterm elections:

2 http://www.washingtonpost.com/wp-srv/politics/special/states/docs/sou96.htm

3 http://www.whitehouse.gov/news/releases/2001/09/20010920-8.html

4 http://www.whitehouse.gov/stateoftheunion/2007/index.html

2007 State of the Union Address

Bipartisanship

Balance Federal Budget

End Earmarks

Fix Medicare/Medicaid

Save Social Security

Reauthorize "No Child"

Affordable Health Care

Immigration Reform

Diversify Energy Supply

Impartial Justice System

Protection from Danger

New Iraq Strategy

Increase Military

Fight HIV/AIDS

Honor Brave Americans

History repeated itself. Like President Clinton, President Bush was trying to be all things to all people. Note the first subject on the laundry list. His effort prompted a comment from another David, David Frum, Bush's own former speechwriter, who called the speech ". . . even more of a hodgepodge than usual."[5]

And, like President Clinton, President Bush (or his speechwriting team) learned a lesson. The very next year, this was the outline for his 2008 State of the Union Address:[6]

5 "Even More of a Hodgepodge Than Usual," by David Frum, *The Wall Street Journal*, January 25, 2007

6 http://www.whitehouse.gov/stateoftheunion/2008/index.html

2008 State of the Union Address

Economy

Education

Trade

Science

Immigration

Iraq/Terrorism

The Flow Structure for this speech was Issues/Actions, the same as Bush had used for most of his other State of the Union Addresses.

Guidelines for Selecting a Flow Structure

When you are selecting a Flow Structure for your presentation, consider these factors:

- **The presenter's individual style.** Choose a Flow Structure that feels right. A little experimentation and practice will help you decide which option works best for you. The fact that one of your colleagues had success with a particular Flow Structure doesn't mean that you will, too. Of course, this mercifully puts to rest the very notion of the one-size-fits-all corporate pitch.
- **The audience's primary interest.** As you've seen, different Flow Structures emphasize different aspects of the story. Choose a Flow Structure that focuses on the interests or concerns of your audience. Remember that Opportunity/Leverage works well for investor presentations and Form/Function for industry peer groups.

  A biotechnology company used the Opportunity/Leverage Flow Structure for their IPO road show by starting with the opportunity (the market for their target disease) and then moving on to how their technology addressed or leveraged that market. Shortly after the company went public, they were invited to appear at an industry conference. For this presentation, they moved their Form (that is, their technology, or their secret sauce) to the top, and moved their Function (that is, the market for their technology) to the later position.

- **Innate story factors.** Some stories lend themselves naturally to a particular Flow Structure. Take advantage of that. For example, a company or industry going through a transition is a prime candidate for the Chronological option.
- **The established agenda.** If you are participating in a conference, seminar, or other gathering in which your presentation is expected to conform to a set format or respond to a particular challenge or question, use a Flow Structure that meets those requirements.
- **Esthetic sense.** Go with your *instinct*. If you have a strong sense that one Flow Structure just "looks good" or "sounds good" or "works well" when applied to your story, use it! Don't try to cram your story into a Flow Structure that feels awkward. If you are comfortable making your presentation, you will transmit your comfort to your audience, and they will empathize with you.

As you can see, selecting a Flow Structure is more of an art than a science. You should also feel free to mix and match Flow Structures. It's quite possible to combine two or three Flow Structures in a single presentation. Try different Flow Structures, but don't limit yourself. *Remember, it's less important which Flow Structure you choose than that you make a choice.*

The Value of Flow Structures

Dr. Emile Loria was the CEO of a French biotechnology company called BioVector Therapeutics, whose science was drug delivery through inhalants. Emile worked diligently with me, over many sessions, to develop the BioVector IPO road show. However, as often happens, the company was acquired instead of going public. Emile then became the CEO of an already-public U.S. biotechnology company called Epimmune, whose science was based on a naturally occurring substance called epitopes. Epimmune had harnessed epitopes to create vaccines to help the body's immune system fight cancer and infectious diseases.

Emile called to ask me to review the Epimmune presentation that he was scheduled to deliver at a biotechnology conference. He emailed me a copy of his presentation. I downloaded it and looked at his overview slide, shown in Figure 4.3.

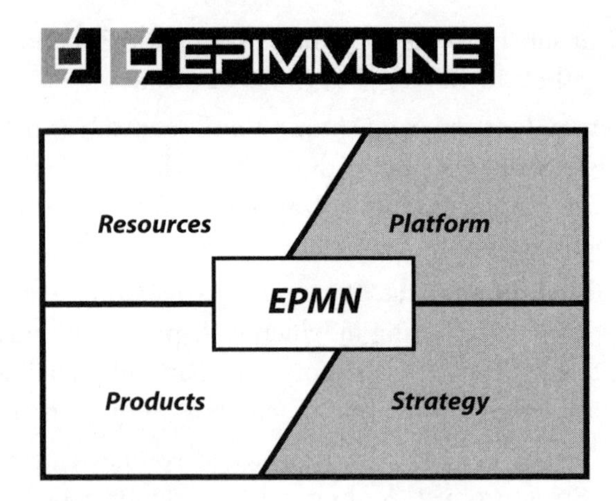

Figure 4.3 *Epimmune overview slide.*

I called Emile and asked him which Flow Structure this represented. Although we had worked together many times, Emile hesitated. But then, after a moment, he said, "*Aha!* Form/Function!"

"Okay. Then why did you start with resources?" I asked.

After a moment, Emile exclaimed, "Of course! I should start with the Epimmune platform, our core science, our Form, and then move on to how it Functions with the products we derive from the platform, our strategy to develop the platform. And I should save the resources until the end to demonstrate how we will implement the strategy!"

"*Aha!*" I concurred.

I spent less than two more minutes shuffling Emile's slides into the correct order (see Figure 4.4), and then I tweaked a couple other slides (very little tweaking was needed for this diligent student). Then I emailed the presentation back to Emile.

It took less than five minutes for us to validate the clarity of Emile's presentation. The forest view.

The Flow Structure approach provides an easy shorthand check of the logic and integrity of your ideas for both you and your audience. Your audience will be able to understand and follow any presentation. Even more important, they'll readily remember your ideas.

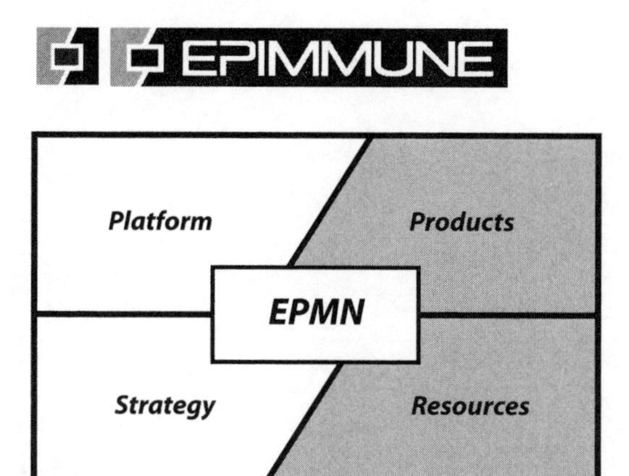

Figure 4.4 *Revised Epimmune overview slide.*

The Four Critical Questions

Let's review everything you've learned so far: Start with the *Framework Form*, do your *Brainstorming* and *Clustering*, and sequence them into a logical path with a specific *Flow Structure*.

All of these steps can be further distilled into the Four Critical Questions:

1. What is your *Point B*?
2. Who is your audience, and what is their *WIIFY*?
3. What are your *Roman Columns*?
4. Why have you put the Roman Columns in a particular order? In other words, which *Flow Structure* have you chosen?

Every time one of my clients asks me to review a presentation, before I look at a single slide or hear a single word, I ask these four questions. Apply the same advice: Pose and answer each of these questions to every presentation you ever give from this moment on.

Your audiences will be grateful, and you will reap the rewards.

Chapter Five

Capturing Your Audience Immediately

COMPANY EXAMPLES

- Intuit Software
- DigitalThink
- Mercer Management
- Cisco Systems
- Yahoo!
- Macromedia
- Argus Insurance
- TheraTech
- Microsoft
- Network Appliance
- Cyrix
- ONI Systems
- Laurel Elementary School

Picture your audience at the start of your presentation. Imagine a group of potential customers who've come to hear you give a presentation about your company's newest product, drifting into a meeting room one by one, sipping their coffee, checking their Blackberrys or I-Phones. Or a banker in a wood-paneled office, sitting behind a tall stack of documents, who must decide whether or not to lend your start-up money for expansion. Or, as with my IPO clients, a roomful of institutional investors in an elegant hotel banquet room, wondering how the NASDAQ is doing at that very moment in time.

Where are their minds? Chances are, not on you, not at the start of your presentation. Chances are they are thinking about an urgent message on their handhelds, the prior loan applicant, the mercurial NASDAQ, their next appointment, the report that's overdue, or the fight they just had with their significant other.

If you were to launch into your presentation at full speed, describing your product, service, or technology, you would vault ahead of your audience, and they would be forced to catch up. *Don't make them think!*

You can capture your audience immediately with an Opening Gambit, a short statement to seize their attention (and simultaneously, to help you launch into your presentation in a comfortable, conversational manner).

In this chapter, you'll learn seven proven options you can use to craft your next Opening Gambit, with examples for each option.

Seven Classic Opening Gambits

1. **Question.** A question directed at the members of the audience.
2. **Factoid.** A striking statistic or little-known fact.
3. **Retrospective/Prospective.** A look backward or forward.
4. **Anecdote.** A short human-interest story.
5. **Quotation.** An endorsement about your business from a respected source.
6. **Aphorism.** A familiar saying.
7. **Analogy.** A comparison between two seemingly unrelated items that helps illuminate a complex, arcane, or obscure topic.

1. The Question

One excellent way to open a presentation is with a question directed at the audience. A well-chosen, relevant question evokes an immediate response, involves the audience, breaks down barriers, and gets the audience thinking about how your message applies to them.

Scott Cook, the founding CEO of Intuit Software, used an Opening Gambit Question to powerful effect. Prior to Intuit's public offering, Scott appeared at the Robertson, Stephens and Company Technology Investment Conference in San Francisco. Here's how he began his presentation:

"Good morning, ladies and gentlemen. Let me begin today's presentation with a question. How many of you balance your checkbooks? May I see a show of hands?" Naturally, almost everybody's hand went up.

"Okay. Now, how many of you *like* doing it?" Everybody's hand went down. There were chuckles in the room, and everyone was listening to Scott Cook.

Scott continued, "You're not alone. Millions of people around the world hate balancing their checkbooks. We at Intuit have developed a simple, easy-to-use, inexpensive new personal finance tool called Quicken that makes balancing checkbooks easy."

It would have been deadly for Scott to launch into his presentation with a detailed description of Quicken; had he done so, it's likely that his investment audience would quickly have fallen into the MEGO state. Instead, his Opening Gambit Question captured their interest and got them involved immediately.

But be careful with the call-for-a-show-of-hands question. It can be considered invasive. Many audiences have been there, done that, and won't appreciate being drawn out that directly. Besides, what if you don't get the show of hands you expect? Yikes! What do you do next? Your question backfires. It worked for Scott Cook because he was a skilled presenter even before starting Intuit, having been a management consultant at Bain and Company.

An effective variation that avoids these dangers of a direct question is to pose a meaningful and relevant rhetorical question to your audience, and then to promptly provide them with an answer. Scott Cook might have said, "If I were to ask how many of you balance your checkbooks, most of you would probably say 'Yes.'"

Mike Pope, who started as the CFO of DigitalThink, Inc., a public company (since acquired by Convergys) that provides custom e-learning courseware, succeeded to the role of CEO. One of Mike's first jobs was to announce a new strategic focus for the company to the employees in an all-hands meeting. He was well aware that the new strategy, which involved belt-tightening, would have an impact on some of the employees' jobs. He chose an approach that emphasized the greater good for the company.

Here's how Mike began his presentation: "If I were to ask Fred here to define DigitalThink's strategic focus, he'd give me one answer. If I were to ask Lisa the same question, I'd get another answer. In fact, I'd probably get as many answers as there are people in the company. You'd all be right. That's because we're trying to be all things to all people. That won't work anymore. I'm here today to define our new strategic focus so that we can all be on the same sheet of music . . . and be successful as a company."

Mike then proceeded to lay out the new strategy. When he concluded his presentation, the audience erupted in spontaneous applause and continued to buzz with positive feedback afterward. By offering them a plan that could produce better results (a WIIFY for them), Mike had prepared them to accept the company's new conditions.

The rhetorical question can be an excellent icebreaker, as long as it's both provocative and relevant to your audience.

2. The Factoid

An Opening Gambit Factoid is a simple, striking statistic or factual statement: a market growth figure, or a detail about an economic, demographic, or social trend with which your audience may not be familiar. This factoid must be closely related to the main themes of your presentation, and to your Point B. The more unusual, striking, and surprising your factoid, the better.

Adrian Slywotzky is a leading management writer and a managing director of Mercer Management Consulting (now merged with and rebranded Oliver Wyman). Adrian often speaks to groups of big-company executives to explain his views about the sources of business growth of the next decade and the strategies business leaders must use to tap those sources. To capture their attention, Adrian begins his presentations with a slide entitled "The Growth Crisis." It lists several of America's biggest, most famous, and most admired companies, from a broad range of industries.

Against this backdrop, Adrian says: "Every investor is looking for companies that can offer consistent double-digit growth in sales and profits. Of all the many great companies listed here, *none* can offer that kind of growth. Not one! It's startling, but true. Our research shows that once you subtract growth from company acquisitions and other special circumstances, *none* of these leading firms has been able to grow at double-digit rates over the past decade."

By the time Adrian has finished communicating this factoid to his audience (which is made up of executives at companies very much like those listed on the slide), many of them are feeling anxious, but all are ready to listen closely to Adrian's suggested solutions to the predicament.

3. The Retrospective/Prospective View

Think of this approach as "That was then; this is now." A retrospective (backward) or prospective (forward) look allows you to grab your audience's attention by moving them in one direction or another, away from their present, immediate concerns: that urgent message on the Blackberry or I-Phone, the mercurial NASDAQ, or that fight with their significant other.

For example, you could refer to the way things used to be done, the way they are done now, and the way you project their being done in the future. The contrast can highlight the value of your company's product or service offerings, thereby framing an effective lead-in to your presentation's main themes and your Point B.

A technology company can quickly capture the evolving power of its solution by showing the speed and capabilities of its products five years ago (a lifetime in the high-tech world), the vastly improved speed and capabilities of today, and the still-greater speed and capabilities of the new line of products that will be going on sale in six months' time.

The Retrospective/Prospective approach can also be used to land a job. I have been privileged to work with Microsoft for most of my 20 years as a coach, and I am often at the company campus in Redmond, Washington. Shortly after the publication of the first edition of this book, a young man came up to me in the corridor carrying a copy and said, "I used a Retrospective/Prospective Opening Gambit to get my job here." I smiled and said, "Tell me about it."

The young man said, "I was in a conference room, being interviewed by several people. Then I stepped up to the front of the room and said, 'Remember when "Yahoo!" was something you said when you were happy? Remember when a web was something spiders spun? And remember when a net was used to catch fish? Well, that was then, and this is now. Now Yahoo! is a major Internet company, the web connects people and computers, and Microsoft's .NET Framework connects those people and computers!'"

The interviewers were so impressed with the young man's story, and that he knew about Microsoft's technology platform for programming applications, they hired him promptly.

4. The Anecdote

By an Opening Gambit Anecdote, I do *not* mean a joke. I like a good joke as much as anyone, but my professional advice to you, and to every single person I coach, is never to tell a joke in a presentation. No one can predict its success or failure. Even if it does get a laugh, in most cases, it will distract from rather than enhance your persuasive message.

An anecdote is a *very* short story, usually one with a human-interest angle. Its effectiveness as an Opening Gambit lies in our natural tendency to be interested in and care about other people. An anecdote creates immediate identity and empathy with your audience. An anecdote is a simple and effective way to make an abstract or potentially boring subject come to life.

Ronald Reagan, "The Great Communicator," never spoke for more than a couple of minutes without using an anecdote to personalize his subject. He was always ready with a brief tale about the brave soldier, the benevolent nurse, or the dignified grandfather as a way of illustrating his themes. Invariably, the tale would coax an empathic nod or a smile of recognition from his audience.

Newspapers and magazines regularly use anecdotes to capture their readers' attention. Just look in today's paper. You're sure to find at least one example of an Anecdote in the opening paragraph. Every professional writer knows how well it works.

Here, from very different settings, are examples of how the anecdote can be used to launch a powerful business presentation:

In the spring of 1996, I worked with Tim Koogle, then the CEO of Yahoo!, the Internet search engine company, as he and his team, including CFO Gary Valenzeula and founder Jerry Yang, planned their IPO road show. Tim and Gary would be the presenters, while the irrepressible Jerry would come along to answer questions. After considering several possible Opening Gambits, Tim decided to begin his presentation with a personal and true-to-life anecdote keyed to a concern he knew he shared with every member of his audience. His anecdote went something like this:

> *Hello, ladies and gentlemen. As you can imagine, going public is a very busy time: There are SEC documents to file, meetings with lawyers and auditors, a road show presentation to prepare, and, of course, a company to run. Imagine how I felt last week when I suddenly realized it was April and I hadn't prepared my tax returns. I had a host of questions about my return, and I hadn't even had a chance to sit down with my accountant.*
>
> *Fortunately, I work for Yahoo!. So I logged on, clicked on the Yahoo! home page, clicked on the menu item called Finance, then clicked on the menu item called Taxes . . . and the answers to all my questions were right there.*

When you consider that Yahoo! provides this kind of powerful Internet search service for a vast array of subjects, from finance to travel to entertainment to sports to health, and then consider the growing legions of users of the Internet, you'll see that the advertising revenues Yahoo! can derive from those legions of users represent a very attractive business opportunity. We invite you to join us.

The Yahoo! IPO story has an interesting extra twist that sheds light on some other presentation principles. A key element of the Yahoo! business model was to promote its brand with a youthful, irreverent, and upbeat image, expressed by its name, its advertising, and even by the cartoonish-like letters of its bright yellow and purple logo. In planning the IPO road show, the Yahoo! team and their advisors pondered ways to capture that brash image without alienating staid investors. At one point, they even considered wearing satin team jackets in the company colors for the road show, and distributing bright yellow kazoos emblazoned with the company logo. These plans were abandoned in favor of a short video intended to set the tone at the very beginning of the presentation. Shot in MTV fashion, replete with fast cuts and odd camera angles, the clip featured exuberant youngsters running, jumping, and performing antics while shouting, "Do *you* Yahoo!?"

By all standards, the video was very well done. And as a television veteran, you might assume that I'd have welcomed the idea of starting a presentation with a video clip. But as I always do when a video is considered, I recommended that it be relegated to second position, *after* Tim Koogle's Opening Gambit Anecdote. After all, I reasoned, the primary investment consideration was management, not slick videos. In fact, Tim had been recruited to run the upstart company because he had *gravitas*. He had served as the president of Intermec Corporation, a division of Litton Industries, and prior to that, had spent eight years at Motorola. Tim's history of success in these respected, "grown-up" businesses would go a long way toward getting the investment community to take the Yahoo! kids seriously. Why not make management the star?

The investment bankers underwriting Yahoo!'s IPO overruled my recommendation. It was probably one of the few times in history that the advice of a media person was rejected by an investment banker as being too conservative. The video ran first. In the end, it didn't matter much. Yahoo! was so hot that its bandwagon rolled merrily along, with legions of eager riders clamoring to get onboard.

Three years earlier, however, in a similar situation, my counsel was heeded by another company, Macromedia, which made software authoring tools for multimedia. Macromedia has since been acquired by Adobe Systems, but at the time, they engaged me to coach their IPO road show team. I didn't get around to starting my program with them until after they had used in-house resources to develop an animated film to open their road show. The film clip was quite impressive. It showed a sparkling, animated, golden M dancing onto the screen, performing spins and leaps to effervescent soundtrack music. Nevertheless, as always, I recommended that the film be relegated to second position, after the CEO's Opening Gambit.

Then I helped Macromedia CEO Bud Colligan develop the following anecdote:

> *Good afternoon, ladies and gentlemen. Welcome to the Macromedia public offering. Last year, I was king of the hill at Apple Computer. I had a great job that provided me with boundless resources, deep staff, and abundant budgets. Why, you might ask, would I make the leap to this risky start-up? The reason is that my job at Apple was to evaluate and develop new technologies. A lot of fascinating new programs and devices came across my desk, but the one that really caught my eye was multimedia.*
>
> *Now I'm sure that you've all heard a lot about multimedia lately. The word seems to be on everybody's lips. But if you were to ask someone to explain the term, most people would have trouble doing so. So rather than try to define multimedia, let me show you multimedia.*

Bud *then* ran the film clip. Thanks to Bud's setup with the Opening Gambit Anecdote, the film was much more than a charming bit of fluff. It was also an impressive demonstration of the kind of creativity that Macromedia could make available to millions of customers, and therefore highly relevant to the Macromedia Point B.

Now let's consider an example from a completely different business setting. Argus Insurance is a Yakima, Washington-based company that sells worksite insurance benefit plans to company employees. These are customized plans that allow each employee to pick the kinds of coverage he or she most needs and wants to pay for, above and beyond the insurance provided by their employer.

Traditionally, Argus agents had simply visited companies, gathered employees in a room, and showed them a generic videotape about their service. The Argus agents then handed out pamphlets and waited to take orders, which were too often few and far between.

When I worked with a group from Argus, I introduced them to the concept of the Opening Gambit. With just a little help, Carol Case, an Argus Insurance agent, came up with the following Opening Gambit based on a real-life scenario:

Last year, one of Argus' customers had a fire in their home. They think an electrical short might have caused it, but no one really knows. Anyway, the home burned down, destroying almost everything they owned. Talk about a disaster! Not only did they lose their home, they were financially crippled as well.

Our insured, like many people, now realizes he could be just one step away from disaster. Luckily, we at Argus Insurance had a solution. Argus reviewed his previous insurance plan and found gaps in the coverage. We then provided quotes for comprehensive, low-cost insurance that provided coverage not provided by his previous policy.

Any human being can identify with this scenario. Carol is signing up many more Argus customers now than in the past, thanks largely to the power of a compelling Opening Gambit.

5. The Quotation

Another option is the Opening Gambit Quotation. That doesn't mean a quotation from William Shakespeare, Winston Churchill, John F. Kennedy, or even Tom Peters . . . unless one of them said something about your company. But if you can provide an endorsement or positive comment about yourself, your products, or your services from *The Wall Street Journal* or the industry press, the quotation provides relevant value. An endorsing quotation can capture your audience's interest and give you credibility at the outset of your presentation.

Please resist the temptation to go to your local bookstore or library and get one of those *Ten Million Quotations for All Occasions* books. Most often, *all* those quotations turn out to be *in*appropriate for *any* occasion.

Please resist the temptation to get one of those Ten Million Quotations for All Occasions *books. Most often,* all *those quotations turn out to be* in*appropriate for* any *occasion.*

When DigitalThink went public in 1999, Mike Pope was the CFO. At the time, the Internet was bubbling into high froth, and a host of other companies in the e-learning space were going public, too. Three years later, upon Mike's succession to CEO, he participated in the Power Presentations program. During the program, Mike smiled knowingly when I introduced the Opening Gambit Quotation option.

Why the smile? Mike explained: "The year we went public, almost every e-learning IPO road show used a quotation from John Chambers [the CEO of Cisco and a kingpin of the Internet]. Chambers said, 'The next killer application for the Internet is going to be education. Education over the Internet is going to be so big it is going to make email usage look like a rounding error.'" The quote was a relevant, validating, and compelling start for any company in that space.

6. The Aphorism

An aphorism, or a familiar saying, can make for an excellent Opening Gambit. But be sure to select one that relates naturally and credibly to your main theme, and to your Point B.

Here is an example of an Opening Gambit Aphorism: A biotechnology company, being formed by the merger of three smaller companies with related sciences in cancer research, launched their presentation with the following: "The whole is greater than the sum of its parts." The original axiom from Euclid, the founder of geometry, is "The whole is *equal to* the sum of its parts," but the biotech presenters gave the familiar saying a little twist. In doing so, they instantly identified the synergies that the new company would enjoy by combining competencies and resources.

Here are a few other examples:

> A company that made graphic display screens used "Seeing is believing" to immediately express the clarity and fidelity of their unique products.
>
> A company with a speech recognition technology used "Easier said than done."
>
> A company that played in a niche market between two giant competitors used "Hit 'em where they ain't." This aphorism, attributed to old-time baseball great Wee Willie Keeler, is a pithy way of saying, "You don't have to be the biggest or the strongest to win in a competition; you can succeed simply by focusing on an area that others are ignoring."

In each case, the Opening Gambit Aphorism triggered the audience's attention at the outset, allowing the presenter to move them into the heart of the presentation. *Grab* and *navigate*.

7. The Analogy

The final option, the Opening Gambit Analogy, is one of the most popular and most effective. An analogy is a comparison between two seemingly unrelated items. In the Introduction to this book, I drew an analogy between a massage therapist and an effective presenter. I hope that got your attention.

A well-crafted analogy is an excellent way of explaining anything that is arcane, obscure, or complicated. If your business deals with products, services, or systems that are technologically complex or that require specialized knowledge to understand, look for a simple analogy that can allow audiences to grasp the essence of the story.

The simpler and clearer the analogy, the better. If you were trying to sell investors on a company that had developed improved software for data network management, you might explain your business by using a twist on a familiar comparison: "Think of us as the people who repair the potholes on the information superhighway. We plan to collect a toll from every driver who travels on our turnpike!"

Although the roadway analogy is a common choice for networking stories, I found an unusual variation of it for the IPO road show of a biotechnology company. TheraTech (now Watson Laboratories, Inc. of Utah) had a technology that enabled controlled-release drug delivery through transdermal patches. During the road show preparations, Charles Ebert, TheraTech's Vice President of Research and Development and a Ph.D. in pharmaceutics, began to discuss the essentials of his core science. As Charles knowledgeably and thoroughly described TheraTech's matrix systems and permeation enhancers, I asked him to pause and explain those terms. Charles said, "Think of the matrix as a truck, think of the drug as its cargo, think of the skin as a border crossing where the truck has to stop, and think of the permeation enhancer as the motor that lifts the barrier and allows the truck to pass through."

Science is complex; by comparison, business is relatively simple. Often, it's difficult to combine the two. While the science behind the sequencing of the human genome is complex, its business potential can be explained through an analogy: "Because our company's researchers were the first to locate and map the genes that cause several major diseases, we'll be in a position to require royalty payments from the big pharmaceutical firms working to eliminate those diseases. It's like owning the copyright on a hit song and collecting a royalty every time it gets played."

Vince Mendillo was the Director of Worldwide Marketing for Microsoft's Mobile Devices when he made the opening presentation at the Mobility Developers Conference in London. Microsoft organized the conference for independent software vendors, or ISVs, in Europe. ISVs write the applications that drive thousands of mobile devices as well as countless other software products.

Vince began his presentation with an Opening Gambit Analogy, made even more effective by leading up to it with this gracious greeting:

> *Good morning! And since I am in Europe, I should add: Bonjour, Buenos dias, and, as an Italian-American, Buon giorno!*
>
> *I'm also a history buff, and I've studied the history of the homeland of my ancestors. For centuries, Italy was divided into many autocratic and warring states. But in 1870, as a result of the valiant efforts of Giuseppe Garibaldi, a guerrilla fighter who became a great general, Italy became a unified nation.*
>
> *The mobile world today is very much like the Italy of yesterday, divided among autocratic warring factions. Now the battle is over technology, platforms, and systems. You ISVs are suffering from the results of that factionalism. We at Microsoft hear you and understand your confusion. In response, we've developed an ecosystem of partners that includes Intel, Texas Instruments, and major carriers, all involved with Microsoft to create an open platform that will enable its members to participate in the deployment of 100 million Microsoft mobile devices. We welcome you here today, and invite you to become part of this ecosystem.*

Vince's Opening Gambit Analogy was gracious, analogous, and unmistakably attention-getting.

Compound Opening Gambits

You can actually combine some of the preceding options for your Opening Gambit. Remember from Chapter 1, "You and Your Audience," how Dan Warmenhoven, the CEO of Network Appliance, began his IPO road show? He started with the line "What's in a name?" (an aphorism: Juliet's immortal query to her Romeo). Then he asked, "What's an appliance?" (a rhetorical question, followed by the answer): "A toaster is an appliance." Then Dan used the toaster as an analogy: "A toaster does one thing and one thing well: It toasts bread."

With this *triple* Opening Gambit, Dan surely had his audience's attention, so he moved on to link the analogy to the rest of his presentation: "Managing data on networks is complicated. It is currently managed by devices that do many things, not all of them well. Our company, Network Appliance, makes a product called a file server. A file server does one thing and does it well: It manages data on networks." Now Dan was ready to move on to his Point B: "When you think of the explosive growth of data in networks, you can see that our file servers are positioned to be a vital part of that growth, and Network Appliance is positioned to grow as a company. *We invite you to join us in that growth.*"

Linking to Point B

To make the opening of your presentation as effective as possible, you need to do more than capture the interest of your audience. The optimal Opening Gambit goes further by linking to your Point B.

In every one of the preceding examples, the presenters continued beyond the Opening Gambit, and then hopped, skipped, and jumped along a path that concluded with Point B. To do the same, you'll need two additional stepping-stones: the *Unique Selling Proposition (USP)* and the *Proof of Concept*.

USP

The USP is a succinct summary of your business, the basic premise that describes what you or your company does, makes, or offers. Think of the USP as the "elevator" version of your presentation: how you'd pitch yourself if you stepped into an elevator and suddenly saw that hot prospect you'd been trying to buttonhole. But please, make it a four-story elevator ride, not a 70-story trip!

The USP should be one or, at most, two sentences long. One of the most common complaints about presentations is "I listened to them for 30 minutes, and I *still* don't know what they do!" The USP is what they do.

Remember Dan Warmenhoven's opening: "Our company, Network Appliance, makes a product called a file server. A file server does one thing and does it well: It manages data on networks." A very clear, concise USP.

Proof of Concept

Proof of Concept is a single telling point that validates your USP. It gives your business instant credibility. The Proof of Concept is optional. Sometimes you can start with the Opening Gambit, link through the USP, and then go directly to Point B without the extra beat.

It is a valuable beat nonetheless, and when you do choose to include the Proof of Concept, you have a number of options. For example, you can use a significant sales achievement: "We sold 85,000 copies of this software the first day it was available"; a prestigious award: "*Business Week* picked our product as one of the top 10 of the year"; or an impressive endorsement: "When an IBM vice president tried our product, he told me he'd give anything if only IBM could have invented it."

Think of your *Opening Gambit*, your *USP*, your *Proof of Concept*, and your *Point B* as a string of connected dynamic inflection points. Once you've segued smoothly through each of them at the start of your presentation, the heart of your argument will be primed for your discussion. You will have *grabbed* your audience's attention, and they will be very clear about what you want them to do.

Let's go back to Scott Cook, the founding CEO of Intuit Software, and the presentation he made at the Robertson, Stephens and Company Technology Investment Conference prior to his public offering. Scott began with two questions: "How many of you balance your checkbooks?" and "How many of you *like* doing it?"

As the chuckling subsided and all the hands dropped, Scott continued, "You're not alone. Millions of people around the world hate balancing their checkbooks. We at Intuit have developed a simple, easy-to-use, inexpensive new personal finance tool called Quicken that makes balancing checkbooks easy." (Intuit's USP.) Now Scott started to roll: "We're confident that those millions of people who hate balancing their checkbooks will buy many units of Quicken . . . " Scott had built up such a head of steam that he skipped past his Proof of Concept, and, *without passing Go*, went directly to his Point B: " . . . making Intuit a company you want to watch."

At that moment in the history of the company, Scott's sole purpose was to raise awareness among potential investors and investment managers of the upcoming Intuit IPO. Unlike the CEOs of other private companies presenting at the conference, all of whom were seeking financing, Scott was there *only* to tee up his road show: " . . . a company you want to watch." Scott was *not* there to raise capital. (In fact, at that very conference, Scott declined a handsome investment offer because he had all the working capital he needed from his primary investors, the Kleiner Perkins Caufield & Byers venture firm.)

As a footnote, Scott is never shy about asking for the order. Of the more than 500 CEOs I've coached for IPO road shows, I've had to prompt almost all of them to make their call to action. (Remember, one of the Five Cardinal Sins is "no clear point.") Most CEOs are reluctant to use the "i" word ("invest"). Far be it from me to suggest that they do. Instead, I recommend words like "invite," "join," "participate," and "share." Only Scott Cook, among those 500, concluded his IPO road show by boldly asking, "Why should you *invest* in Intuit?" and then providing them with the answers.

Remember from Chapter 4, "Finding Your Flow," how CEO Jerry Rogers began the Cyrix IPO road show: "Cyrix competes against the established giants Intel and AMD, as well as two other large, well-funded companies." (Jerry's Opening Gambit was a factoid . . . in this case, a *most* striking statement.) "So as a tiny start-up trying to establish itself, I have to respond to some challenging questions from potential new customers about the IBM-compatible microprocessors that Cyrix designs." (Cyrix's USP.) "Three questions keep recurring: 'Will Cyrix microprocessors run all software applications?' 'How will Cyrix compete with Intel?' and 'Does Cyrix have the financial stability to succeed?'"

Jerry continued, "Cyrix has already begun shipments of the first commercially available 486 microprocessors *not* produced by Intel." (Cyrix's Proof of Concept.) "But those same three questions about compatibility, competition, and finances remain important to potential investors like you. In my presentation today, I'll provide you with the answers to those questions to demonstrate that Cyrix is indeed a sound investment." (Cyrix's Point B.)

Jerry's answers and his call to action proved to be effective: When Cyrix launched as a public company, their shares were priced at $13, and the first trade was over $19, a significant achievement in 1993.

Now let's move from CEO Jerry Rogers to Carol Case, the salesperson from Argus Insurance you met earlier in this chapter. Carol began her pitches to her prospective buyers with the scenario about the fire. After the scenario, Carol continued: "Our customer is like many customers who purchase a basic policy not customized to their individual needs. He now realizes that means being just one step away from disaster. Luckily, we at Argus Insurance have a solution. Argus can provide you with a customized, value-added package of insurance that provides for your individual needs to protect you against serious financial loss." (Argus' USP.) Carol was already teeing up her Point B, but for good measure, she added her Proof of Concept: "Maybe that's why Argus is one of the fastest-growing insurance brokers in the state." Now, she was ready for an even stronger Point B: "I know that you'll want to take advantage of the opportunity and sign up for this important coverage today."

When you power-launch your presentation with your *Opening Gambit*, your *USP*, your *Proof of Concept*, and your *Point B*, your audience will have no doubt about where they're going. Now it's time for you to tell them how you intend to *navigate* them there.

Tell 'em What You're Gonna Tell 'em

The prior dynamic inflection points will prime your audience, but do you want to now dive directly into the body of your presentation? Do you want to start drilling deeply into the first of the clusters you selected? Not quite yet. I suggest that first you take a moment to give your audience a *preview* of the outline of your major ideas.

The technique for helping your audience become oriented and track the flow of your ideas is the classic *Tell 'em what you're gonna tell 'em*. You can also think of it as the *forest view*.

In most business presentations, this preview is expressed in the Overview or Agenda slide. In an IPO road show, it's in the Investment Highlights slide. In either case, it summarizes the chief attractions of a company's offering. Why not map it to the sequence of the Roman Columns? Why not make it track the entire presentation? In that way, you and your audience can see all the major Roman Columns and the Flow Structure that unifies them. This is a view of both the trees and the forest. Think of it as a final quality check.

But telling your audience what you're gonna tell 'em has much more to offer than an agenda. You can also extend your narrative string with two more dynamic inflection points: *link forward from Point B* and *forecast the running time of your presentation*.

Link Forward from Point B

In the Argus Insurance sales pitch, after Carol Case asked for the order with "I know that you'll want to take advantage of the opportunity and sign up for this important coverage today," she continued on to her Overview slide with " . . . so that you can consider signing with Argus . . . ", and then she ran through the bullets of her Overview slide.

Contrast Carol's forward link with the far more common approach of *no* link at all: "Now I'd like to talk about my agenda . . . "

In the opening of the Intuit IPO road show presentation, after Scott Cook's call to action with "Why should you invest in Intuit?", he made a *forward link* to his Investment Opportunity slide (shown in Figure 5.1) with "These are the reasons to consider Intuit for your portfolio." Then he ran through the bullets, briefly, adding value to each point.

In doing so, Scott avoided the all-too-common practice of reading the bullets verbatim; a practice that invariably annoys *any* audience, because they think to themselves: "I'm not a child! I can read it myself!"

Investment Opportunity

- **Substantial, underpenetrated markets**
  - **Personal finance**
  - **Small business accounting**
  - **International**
  - **Services**
- **Proven consumer products approach**
- **Best of breed products**
- **Market share leader across all platforms**
- **Significant revenue from existing customers**

Intuit

Figure 5.1 *The Intuit IPO road show Investment Opportunity slide.*

When Scott was done with the bullets, he added, "Please consider this the outline for the next 20 minutes of our presentation. Let's start at the top with the market opportunity," adding *yet another* forward link, this time into the heart of his presentation.

Forecast the Time

Instead of plunging your audience headlong into a dark tunnel, establish the endpoint at the beginning of your presentation; show them the light at the end *while they are still at the entrance*. By stating how long your presentation will take, you demonstrate that you respect the value of your audience's time and intend to use it productively. Another aspect of Audience Advocacy.

By providing your audience with a *road map* and a *forecast* of the time, you are giving them a *plan* and a *schedule*. These four italicized nouns have a least-common-denominator noun: *management*. Once again, you're sending the subliminal message of *Effective Management*.

Of course, just because you've told 'em what you're gonna tell 'em, no one will think of you as an excellent manager. That's a stretch. But the converse proves the point: If you plunge your audience headlong into that dark tunnel without providing a road map or endpoint, they'll feel out of control. If instead you show them the light, they'll sit back, lock into your message, and subconsciously say to themselves, "These people know what they're doing. They have a plan. They're well prepared. Let's listen to what they have to say."

Telling 'em what you're gonna tell 'em also provides additional benefits to you, the presenter. When you click on your Overview slide and run through it for your audience, you can mentally check the components of your presentation and remind yourself of the overall flow of ideas. This will make it easier for you to move confidently from point to point as you step through the body of your presentation. Remember how well that worked for Hugh Martin of ONI Systems and his "Top 10 Questions from Institutional Investors"?

Furthermore, it's quite likely that someday, someone will come up to you five minutes before your presentation and say, "Gee, I'm sorry, but we're running awfully late. We planned to give you half an hour, but we can only spare eight minutes."

Panic time? No; simply move to your Overview slide and walk your audience through your key points in the allotted time. After all, since the Overview slide is a miniature view of your entire presentation, you can use it to support an ultra-brief condensation of all your ideas.

That old chestnut *Tell 'em what you're gonna tell 'em* has a host of benefits: road map, linkage, forecast, prompt, and short form.

You can now proceed to fulfill your forecast by telling 'em what you promised. That is, move through each of your clusters in greater detail, putting flesh on the sturdy bones of your structure.

Of course, when you're done with all the clusters, you then tell 'em what you told 'em. In most presentations, this is in the Summary slide; in an IPO road show, it's in the Investment Highlights slide. In both cases, it's a mirror of its predecessor from the beginning of the presentation. This mirroring, this linking of the preview and the summary, serves to bookend the presentation. This also creates a thematic resolution that neatly ties together the entire presentation. Express that resolution with a final restatement of your Point B. The last words your audience should hear is your call to action.

The last words your audience should hear is your call to action.

An important footnote about your summary: Make it brief! In a book, when only a few pages are left and the reader knows that the end is in sight, the writer quickens the pace of the narrative. So it is with a presentation. When you get to your Summary slide, discuss it succinctly.

You have now effectively *grabbed* your audience at the beginning, *navigated* them through all the parts, and *deposited* them at Point B. Moreover, your whole story has now fulfilled Aristotle's classical requirements for any story: a strong beginning, a solid middle, and a decisive end.

In the Power Presentations programs, we provide a Story Form (see Figure 5.2) that captures all the preceding points. It's a useful tool for tracking a basic format that will work for any persuasive presentation.

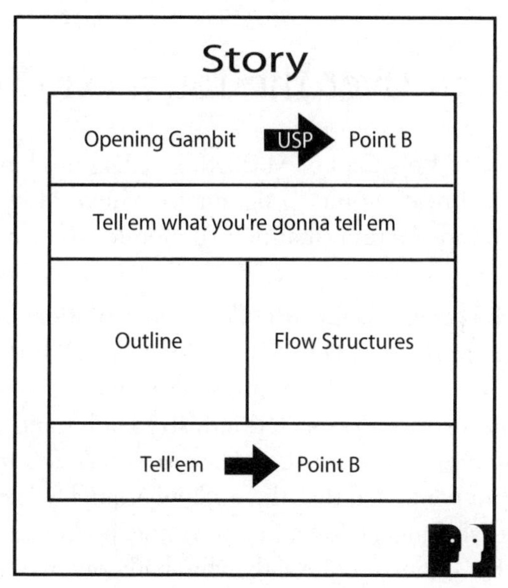

Figure 5.2 *The Power Presentations Story Form. (You can download a copy of this form by visiting our website, www.powerltd.com.)*

90 Seconds to Launch

The *Opening Gambit*, the *USP*, a *forward link* to *Point B*, another *forward link* to *Tell 'em*, and one more *forward link* into the *Overview* are the first important steps of *any* presentation. This continuous string creates a dynamic thrust that propels and energizes the balance of your time in front of your audience.

It's important that you hit these points, and in that order, within the first 90 seconds of your presentation . . . at the maximum. Always remember the importance of the start of your presentation. If you lose your audience within that first

90 seconds, chances are they will be lost forever. Remember, you never get a second chance to make a first impression.

You never get a second chance to make a first impression.

But if you capture your audience's attention, define your Point B, and establish your credibility in those same 90 seconds, the audience is yours, and they will follow you wherever you want them to go.

Winning Over the Toughest Crowd

Jim Flautt was the Vice President of Marketing at DigitalThink. He participated in the same Power Presentations program in which Mike Pope, the newly ascended CEO, developed a presentation to announce the company's new strategic focus.

The day after the program, Jim attended a presentation at the alumni association of his graduate business school. He decided to watch from the back of the room, as I do, to observe not only the presenter, but the audience reactions as well. The guest speaker was a respected industry leader and a frequent presenter, but his presentation was, from the outset, an unabated torrent of excessive words and slides. Jim could see that the audience quickly lost interest: MEGO. As he watched the shifting heads and squirming bodies in front of him, Jim smiled knowingly. He understood exactly what had gone wrong.

The next day, Jim, who happens to be an Annapolis graduate and a former naval officer aboard the *USS Albany*, was scheduled to speak about submarines to his son's first-grade class at Laurel Elementary School. Jim realized that a roomful of seven-year-olds, who represent the ultimate in short attention spans, would be as challenging as any business audience. So he set about to make use of what he'd learned about making business presentations to prepare for the first-graders.

In the style of Intuit's Scott Cook, Jim began with a call-for-a-show-of-hands Opening Gambit, a format quite familiar to his young audience. Jim had three questions: "How many of you know what a submarine is?" All the first-graders raised their hands. "How many of you have ever seen a submarine?" Half the first-graders raised their hands. "How many of you have ever seen a submarine *fly*?" Now all the first-graders smiled, giggled, and gasped. Jim had their attention.

Jim continued on to his *Point B*: "Well, I'm here today to tell you all about submarines." Then, to *link forward* from his *Point B*, Jim *told 'em what he was gonna tell 'em*: " . . . what goes on in a submarine, how to drive a submarine, how to repair a submarine, and some cool things that a submarine does. When I'm done with all of that, about 15 minutes from now . . . " (Jim's *forecast for time*) " . . . I'll *show* you a submarine that flies!" Just for good measure, Jim added a big WIIFY.

This got more smiles, giggles, gasps, and now, delighted cheers from the children. Jim had them and he kept them, with simple slide photographs, in rapt attention for about 15 minutes, until the very end. Then, true to his promise, Jim clicked on his computer and ran a video clip of a U.S. Navy submarine going through an emergency surfacing exercise, leaping out of the roiling ocean looking, for all the world, like a dolphin performing a stunt.

Jim did it all: he grabbed, navigated, and deposited. If it can work with wired seven-year-olds, it can work with occupationally-short-attention-span investors looking to make a return on their investment, with a concerned customer looking to find a product that performs a vital function better than the current solution, or with a stressed manager looking for a strategy to compete more effectively. It can work for you.

Chapter Six

Communicating Visually

COMPANY EXAMPLE:

- **Microsoft**

The Proper Role of Graphics

Think about a time when you were in the audience at a presentation and the graphics didn't work. What was the problem? These are the most common answers my business clients give:

- "The graphics were cluttered."
- "There was too much on the slide."
- "The slide looked like an eye chart."
- "The slide was a Data Dump."

Now flip the lens and take the point of view of an Audience Advocate. What's the effect on you? Odds are that it's another case of the dreaded MEGO syndrome: the same cause and effect as when a story is unloaded on you as a Data Dump.

The main reason this happens is that presenters fail to distinguish between a document and a presentation. They treat a presentation as a document. This is the *Presentation-as-Document Syndrome*, in which the presenter uses the graphics as both a display and as a record, as both show and tell. As you read in the Foreword, this is the vestigial legacy of the ancient origins of presentations, the flip chart.

Business documents include:

- *Annual reports*, filled with dense text and highly detailed tables, charts, and graphs
- *Strategic plans*, filled with dense text and highly detailed tables, charts, and graphs
- *Market analyses*, filled with dense text and highly detailed tables, charts, and graphs
- *Meeting notes*, filled with dense text and highly detailed tables, charts, and graphs

See the pattern? All these types of documents are necessary and important in their place, but business documents are *not* presentations.

So the true problem with presentation graphics is that, all too often, presenters take a flood of data, those dense text and highly detailed tables, charts, and graphs, and simply reproduce them, with little or no modification, for their presentation graphics.

This *Presentation-as-Document Syndrome* represents one of the most common problems that plagues presentations. Presenters have become so accustomed to relying on graphics, especially Microsoft PowerPoint slides, that they often think of the presentation as a mere accompaniment to those aids. In fact, many people act as if the presentation is completely dispensable. They'll say, "I can't attend your presentation next week. But it doesn't matter. Just send me your slides!" Or they sometimes say, "Send me your slides in advance." The PowerPoint slides then are treated as *handouts*.

What's more, presenters frequently provide the handouts to their audience before the presentation. The audience then *reads* the handout, *sees* the slide, and *hears* the presenter read what is on the slide. This is known as "triple delivery," an assault on the audience's senses that leads to that lethal MEGO. Using slides as handouts is but one manifestation of the Presentation-as-Document Syndrome. There are three others.

- Use the slides as *notes* to help the presenter remember what to say.
- Cram a plethora of *details* on the slides, as if to demonstrate legitimacy.
- Fill the slides with enough information so that anyone else in the company using the same slides will maintain the *uniformity* of the message.

But the name of my company is *not*:

- Power Handouts
- Power Notes

- Power Details
- Power Uniformity

It is Power Presentations.

A presentation is a pure play. It must serve only one purpose. Remember the words of Dan Warmenhoven, the CEO of Network Appliance, in the Opening Gambit of his IPO road show: "Do one thing and one thing well." If a presentation tries to serve two or more purposes, it dilutes both purposes. The presentation itself is neither fish nor fowl.

A presentation is a presentation and *only* a presentation . . . *never* a document. After all, Microsoft provides Word for documents and PowerPoint for presentations. *And never the twain shall meet.*

A presentation is a presentation and only *a presentation . . .*
never *a document.*

If you *do* need a document of your presentation, Microsoft PowerPoint provides the Notes Page view, as represented in Figure 6.1. The top of the Notes Page contains only what your audience sees projected on the screen. The bottom provides the additional material for the handouts.

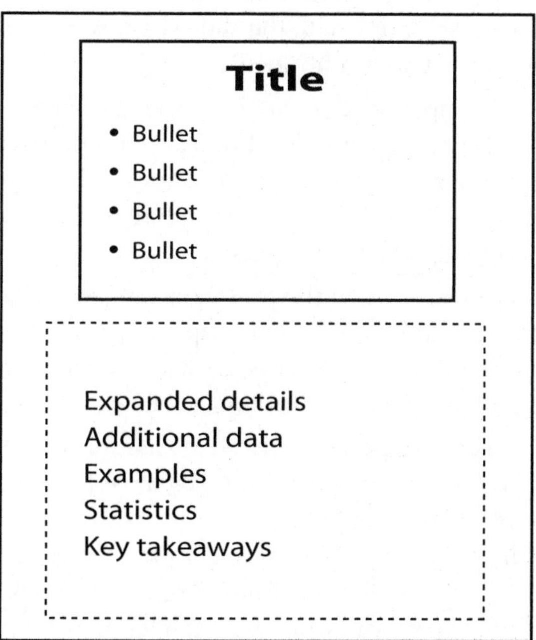

Figure 6.1 *The Notes Page view in Microsoft PowerPoint.*

Be sure to distribute the handouts only *after* the presentation. If you distribute them before or during the presentation, your audience members will flip through them as you speak, and they won't listen to what you have to say.

If you're asked to provide a copy of your presentation for a conference so that the slides can be printed in book form, use PowerPoint's Notes Page view. That way, you'll maintain the integrity of your slides as purely presentation material.

If you're asked to provide a copy of the presentation in advance, as so often happens, especially in the venture capital and financial sector, politely offer to provide a business plan or executive summary . . . as a document. And create that document with Microsoft Word, *not* PowerPoint.

Presenter Focus

An extension of the Presentation-as-Document Syndrome is what happens to the audience when the screen lights up with a slide filled with dense text and highly detailed tables, charts, and graphs. The focus of the audience immediately, *and involuntarily*, goes to the graphics, and they start to read. When they start reading, they stop listening. The graphics then become the center of attention, and the presenter becomes subordinate to the slide show, serving, at best, as a voice-over narrator and, at worst, as a ventriloquist.

This problem is compounded as the presenter becomes a reader, too. The reading often fails to rise above the level of a verbatim recitation. Reciting the slides verbatim is patronizing to the audience. They think to themselves: "I'm not a child! I can read it myself!" The results are a failure to connect, a failure to communicate, and, most likely, a failure to persuade.

An even worse variant is when the presenter rambles, talking about subjects that have nothing to do with the slide. This disparity jams the audience's eyes and ears, producing confusion and annoyance. It's like crossing a DVD player's video and audio cables, producing a scrambled image and static.

How, then, should you approach effective slide design? View graphics the way my clients at Microsoft view them. The Executive Presentations Team is the specialized unit at Microsoft charged with producing the slides and graphics necessary to support the senior executives at major presentation events for the company. It was formerly headed by Jon Bromberg, an alumnus of the New York theater world. After 11 years with Microsoft, Jon has now retired, but during his tenure he was the director of events, in charge of a huge operation with state-of-the-art production and graphics capabilities.

Despite all those technical bells and whistles, the Executive Presentations Team refers to PowerPoint, Microsoft's own product, as *speaker support*. The slides or other graphics are there to support the presenter, not the other way around.

The slides or other graphics are there to support the presenter, not the other way around.

This is the very same model that we see every evening on the television news broadcasts. We see the anchor: Charles Gibson, Brian Williams, or Anderson Cooper, giving us information they interpret for us. Yes, they use graphics, but their graphics play a supporting role. The images often consist of little more than a simple picture and a word or two to headline the story that the anchor is presenting: a photograph of the Capitol building and the words "Tax Proposal," or a picture of a medicine bottle and the words "Prescription Drugs." Gibson, Williams, or Cooper is always center stage.

It so happens that my cousin, Joel Goldberg, has worked for more than 25 years as a graphic artist at ABC News. Using elaborate electronic tools, Joel often spends an hour getting the appearance of a single image or word exactly right, with highlights, tints, dimension, shadowing, and other tricks of his trade. That image or word may appear on the screen behind Charles Gibson for just a few seconds before vanishing forever. While I'm sure that Joel is paid well for his talents, it is Charles Gibson who pulls down a much larger salary for the gravitas he brings to the news.

This Presenter Focus/Graphics Support relationship is the only effective model for a presentation. The presentation cannot serve as a document unless it is complete in itself, in which case the audience wouldn't need the presenter. They could sit in silence and read the slides to themselves. On the other hand, if the graphics constitute a *partial* document, they cannot stand alone, and they serve only to distract attention from the presenter.

Instead, when the presenter interprets for the audience and the graphics provide support, the presenter can lead the audience to a conclusion. When this happens, the presenter manages the audience's minds, creating the subliminal takeaway: *Effective Management*.

The audience also takes away a visual reinforcement of the presenter's message. As an ancient Chinese proverb tells us:

I hear and I forget;

I see and I remember;

I do and I understand.

Less Is More

To make all these things happen, we need a guiding principle. That principle is *Less Is More*. These words have been attributed to one of the foremost architects and designers of the 20th century, Ludwig Mies van der Rohe (1886–1969), the father of the minimalist school. Mies directed the influential Bauhaus School of Design in Germany in the 1930s and then came to the United States, where he designed such sleek, classic structures as the bronze-and-glass Seagram Building in New York City.

Mies' famous Less Is More dictum became the guiding principle for many of the greatest designers of the past hundred years. Less Is More should be your guiding principle when you are creating your presentation graphics.

It is mine. It is mine after years of working in television, with access to the vast capabilities of professional graphic artists like my cousin Joel, who operate in multimillion-dollar electronic control rooms known as "paint boxes." It is mine after 20 years of working with evolving generations of computer-based graphics programs, culminating in the latest version of Microsoft's PowerPoint, whose capabilities approach those of the professional paint boxes. These powerful tools notwithstanding, when designing graphics, I rely on the wisdom of Less Is More, and its corollary, *When in doubt, leave it out*.

An important benefit of a slide designed with this minimalist approach is that it serves as an instant prompt for the presenter: a visual mnemonic.

Perception Psychology

In addition to Presenter Focus and Less Is More, the two essential concepts for powerful graphics design, there is a third vital element in the equation: the audience and how they take in what they see. This is *Perception Psychology*.

Let's begin by analyzing how the human eye moves.

You're familiar with the illuminated initial letter that appears in fine manuscripts or in handsomely printed books. It always appears in the upper-left corner of the page, where a new book or a new chapter begins. Magazines and newspapers also often use an enlarged initial letter to start articles. That's because in Western culture, countries like the U.S., France, Spain, Italy, and Germany, languages are always printed from left to right and top to bottom. As a result, a Western reader's eyes are conditioned to move to the upper-left corner to start a new passage.

This movement is not innate; it is culturally determined and learned. In the Middle East, books begin at the back and are read from right to left. Therefore, the eyes of those readers are conditioned to begin at the upper-right corner of a page.

Think of this tendency for our eyes to jump to the upper-left corner of a page as the *conditioned carriage return*, since the movement recalls the repeated movement of the carriage on an old-fashioned typewriter.

When people read print in a book, magazine, or newspaper, their eyes make the conditioned carriage return with every new page. In a presentation, this movement occurs with every new slide.

There is one major difference, however. When your eyes shuttle across a page in a book or magazine, they move only five to eight inches at a time. In a presentation, when they have to leap across a large screen in a conference room or auditorium, they move anywhere from two feet to 20 feet, depending on the size of the screen.

Therefore, every time you click to a new graphic on the screen, your audience's eyes are driven by two powerful and contradictory forces: First, because of a lifetime of conditioning, they jump to the upper-left corner. Then their eyes suddenly become aware that there is more information on the screen, and so now their eyes must make another move to take in the rest of the information. This next move is more powerful than the first. Their eyes sweep to the right in a completely involuntary action. This is the *reflexive cross sweep*, illustrated in Figure 6.2.

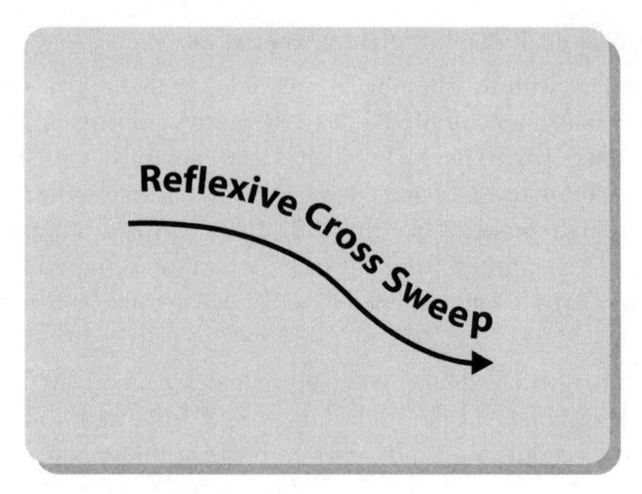

Figure 6.2 *The reflexive cross sweep.*

Unlike the learned move of the conditioned carriage return, the left-to-right movement of the reflexive cross sweep is involuntary. Sometimes the move is down and to the right, and sometimes it is up and to the right. Most painters organize their pictures based on eye movement down and to the right. That is why, more often than not, artists sign their canvases in the lower-right corner.

Businesspeople also intuitively follow the reflexive move to the right. But they are also accustomed to taking the high road, following the desired growth pattern of revenue and profits symbolized by the hockey stick (see Figure 6.3).

Figure 6.3 *The hockey stick movement, up and to the right.*

Whether up and to the right (as in business) or down and to the right (as in art), this involuntary left-to-right movement is deeply embedded in our nature.

No one has ever fully explained why these reflexive movements occur. In his 1954 book, *Art and Visual Perception: A Psychology of the Creative Eye*, Rudolf Arnheim, a Gestalt psychologist and a scholar of art and cinema, offered several theories, ranging from the plausible to the fanciful. The plausible: Because the majority of humans are right-handed, there is a natural tendency toward greater awareness of objects on the right side of the visual field. The fanciful: Early homo sapiens were so impressed by the movement of the sun from left to right that they favored this form of movement

All these theories notwithstanding, the innate predisposition of the human eye to move from left to right is irresistible. You can feel it yourself as you scan this very page or any of the illustrations in this book.

Television and cinema directors incorporate this preference in how they direct their subjects and cameras. Next time you watch a well-directed movie or television drama, notice how the characters move across the screen. Most often, the sympathetic characters, the heroes and heroines with whom the audience identifies, move from the left side of the screen toward the right, flowing with the natural movement of the eye. By contrast, the unsympathetic characters, the villains whom the audience dislikes, move from right to left, fighting the eye's natural flow.

Even when the characters are stationary, the movement of the camera can convey the same feelings: A pan right is smooth; a pan left drags. Very subtly, these differences fuel the audience's emotional reaction to the drama, helping them to think and feel the way the actors, director, and writer intended.

Director Sam Mendes used these techniques to powerful effect in the Tom Hanks/Paul Newman film, *The Road to Perdition*. In the opening scene, we see a city street in the Depression era crowded with pedestrians, most of them moving from left to right. Then a young boy on a bicycle enters, pedaling from right to left, against the grain. The boy is Tom Hanks' son, whose difficulties form the central part of the story. From the outset, these powerful cinematic dynamics set the foreboding tone for the rest of the film.

In the theatre, directors incorporate the same approach with actors on the stage: protagonists move toward the right, and antagonists move toward the left.

It all goes back to how we first learned to absorb information as children: reading text. By understanding Perception Psychology and applying it properly, you can control the effect of your graphics on your audience. And in presentations, you want that effect to be positive.

By understanding Perception Psychology and applying it properly, you can control the effect of your graphics on your audience.

In a presentation, when a new image flashes on the screen, your audience's eyes will make two instant moves: one to the left to start the slide (the conditioned carriage return), and one to the right to take in all the information (the reflexive cross sweep). The move to the right will either up or down.

However, if, in the design of your graphics, you've put excess data on the screen, your audience cannot take in the entire image in two moves. They're forced to make another move, and perhaps more than one. This third move, along with any subsequent moves they must make, will be hard work, backwards and against the grain. This third trip, back and to the left, is a *forced carriage return*. You can see all three moves in Figure 6.4.

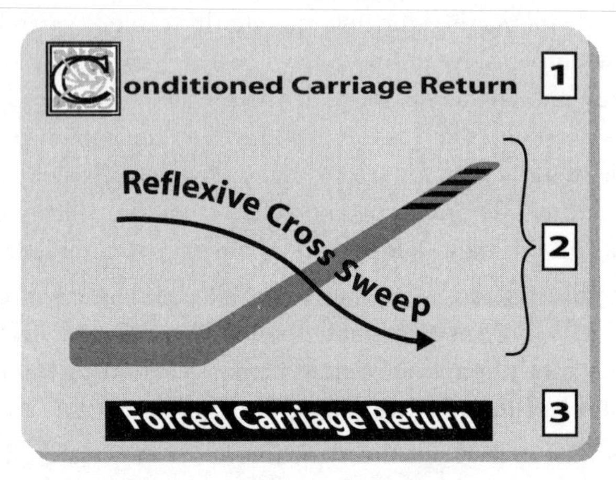

Figure 6.4 *Three moves caused by excessive graphic information.*

This again leads us to one of the most important practical applications of the Less Is More dictum: *Don't make me think!* That refrain asks you not to make your audience work to understand your ideas. The same refrain applies to the work your audience must do to absorb your graphics. Therefore, design all your slides to *Minimize the Eye Sweeps* of your audience. For *every* graphic, keep the number of times their eyes must go back and forth across the screen to an absolute minimum. Make it easy for your audience, and they will make it easy for you. Think of the alternative!

Design all your slides to Minimize the Eye Sweeps *of your audience.*

Thus, the overarching principles of powerful and effective graphics are

- Presenter Focus
- Less Is More
- Minimize Eye Sweeps

Keep these principles uppermost in your mind and apply them to all your graphic design elements.

Graphic Design Elements

All the bells and whistles in all the graphics programs available for business presentations boil down to just four basic design elements:

- **Pictorial.** Photographs, sketches, maps, icons, logos, screen shots, or clip art.

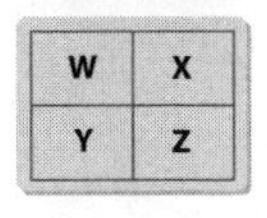

- **Relational.** Tables, matrices, hierarchies, and organizational charts. (Think of a relational chart as an image that captures a set of relationships, connections, or links in visual form.)

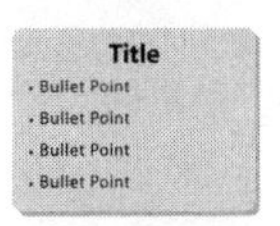

- **Text.** Text comes in two flavors: bullets and sentences.

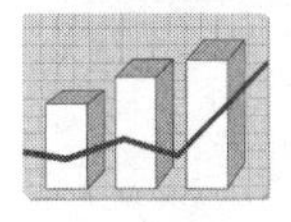

- **Numeric.** Numbers expressed in bar charts, pie charts, area charts, line charts, histograms, and other, more specialized types of graphs.

When you choose either a pictorial or a relational graphic, you achieve Less Is More by default. *A picture is worth a thousand words.* And a table or chart, by definition, organizes multiple diverse elements, turning more into less. So both pictorial and relational graphics promptly fulfill Mies van der Rohe's principle.

However, much business information can't be captured in a picture or a table. As a result, most presentations bulk up with text and numeric graphics. This is where the trouble begins. In the first place, presenters end up using more, not less, on the screen, thereby losing effectiveness. What's worse, they fall prey to the Presentation-as-Document Syndrome.

A presentation is a presentation and only a presentation. The primary role of graphics is to support the presenter and to give the presenter the opportunity to add value above and beyond what is projected on the screen.

Graphics also help the audience remember your content. Remember the Chinese proverb: "I see and I remember."

In the next chapter, you'll learn how to create effective text graphics; and in the one after that, you'll learn how to create effective numeric graphics. The emphasis in both of these chapters will be on how to achieve Less Is More, thereby conveying your visual information in a clear, crisp, and powerful form.

Chapter Seven

Making the Text Talk

Bullets Versus Sentences

In the previous chapter, you saw that all text slides come in only two options: bullets and sentences. Each of these options is quite different, with separate forms and functions. Keep them distinct.

A bullet is meant to express a core idea, so craft it in the form of a headline. Look at any newspaper, and you'll see that a headline is not a complete sentence. Basic English grammar dictates that a sentence must contain a subject and a verb, but most headlines are not complete sentences. Generally, headlines omit the parts of speech that form complete sentences: articles (the, an, a), conjunctions (and, but, or), and prepositions (of, for, by, through).

Why are headlines written in this shorthand style? There are several good reasons. When fewer words have to be squeezed into an available space, the size of the letters can be increased, enhancing legibility. Furthermore, by providing the gist of the story in a few words, readers can scan a page full of stories in a few seconds and pick out the ones of interest.

Legibility and speed are equally important in presentation slides. When you create a text slide containing bullets, you are, in effect, presenting headlines only. Where does the body text appear? Not on any slide. As the presenter, it is *your* job to put flesh on the bones of the headline bullets. The presenter provides the body text. The presenter is the focus of the presentation.

The presenter is the focus of the presentation.

This approach can make for a very crisp, clear presentation. You can summarize most of the concepts of your story (distilled in the Brainstorming process and organized into clusters) in two-to-five-word headline-style bullets. Some typical concepts from any company story might include

- Breakthrough New Product Line
- Experienced Management Team
- Exploding Market
- Targeted Strategy

How long would you be able to speak about any of these concepts as they apply to your business? Probably for several minutes each, if not longer. Therefore, the optimal presentation is composed of a presenter providing spoken body text for headline-style bullets on the slides.

What about sentences? When should you use them in your graphics?

The only time you need a sentence is when you need to demonstrate verbatim accuracy. Use a sentence only when you're citing the specific words in a quotation, like this:

> "PQR Technologies is the most exciting new business concept I've seen this year."
>
> Tom Hudson
> *High-Tech Monthly*

While you can use full sentences for your endorsing quotations, you would do well to keep them to a minimum, and rely primarily on bullets as headlines for your text slides.

An important underlying reason for avoiding sentences on slides goes back to the basic principle of Audience Advocacy. Sentences are longer than bullets, and they usually extend over several lines. Therefore, reading them on the slide generally requires additional eye sweeps, making your audience work harder to absorb your message. Make it easy for the audience to take in your graphics. *Minimize Eye Sweeps.*

Wordwrap

When a bullet is too long to fit on a single line, the text automatically continues on to a second line. This is called wordwrap, and it looks like Figure 7.1.

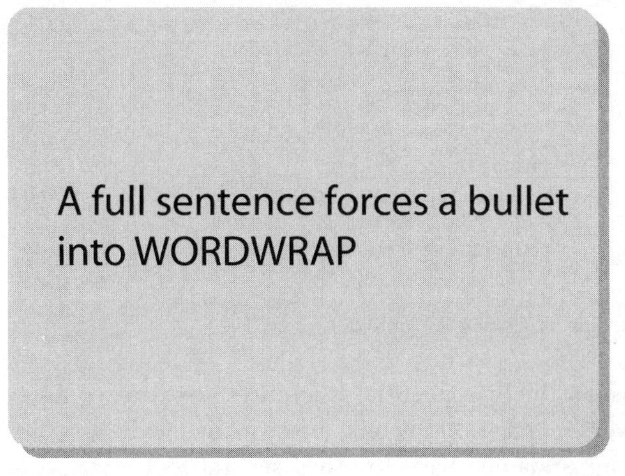

Figure 7.1 *Wordwrap forces two eye sweeps.*

A full sentence almost always causes wordwrap, which requires an extra eye sweep . . . and more work for the audience. Shift the sentence-style bullet to a headline-style bullet by omitting articles, conjunctions, prepositions, and other unnecessary words. Minimize Eye Sweeps by restricting the bullet to a single line.

Crafting the Effective Bullet Slide

You'll recall that I use the Problem/Solution Flow Structure in my programs and in this book. Most of the previous chapters have begun by asking you to think about presentation problems. In the same spirit, let's look at the particularly problematic bullet slide shown in Figure 7.2.

Does this look familiar? When I show this slide during my program, the groans and laughter from the participants make it clear that slides like this are all too common. This is a true Data Dump slide, one that is in desperate need of surgery.

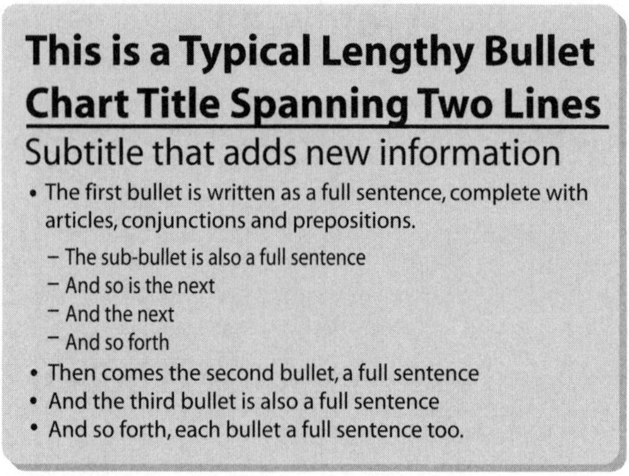

This is a Typical Lengthy Bullet Chart Title Spanning Two Lines

Subtitle that adds new information

- The first bullet is written as a full sentence, complete with articles, conjunctions and prepositions.
  - The sub-bullet is also a full sentence
  - And so is the next
  - And the next
  - And so forth
- Then comes the second bullet, a full sentence
- And the third bullet is also a full sentence
- And so forth, each bullet a full sentence too.

Figure 7.2 *How not to design a bullet slide.*

First, consider the two-line title. It requires an extra eye sweep. In this book, your eyes traverse a page that is only a few inches wide. In your presentations, the eyes of your audience have to travel several feet across a large projection screen. The more you make their eyes travel, the harder they have to work. Be an Audience Advocate! Replace the two-line title with a one-liner that states a single concept.

Next, notice that the subtitle, which adds new information, makes the audience do extra work not only to interpret the new information, but also to determine its relationship to the main title. Instead, replace it with a subtitle that relates easily and obviously back to the main title. Better still, eliminate the subtitle altogether. *Less Is More.*

Next, consider the first bullet. It has all the parts of speech that make a bullet into a sentence. This gives the audience more work to do by adding extra eye sweeps. Simplify the bullet into a newspaper-style headline. Do the same with all the other bullets and with the sub-bullets as well.

Now for a subtle, but important, point: The default in most graphics and word processing programs is to indicate a sub-bullet with a dash. This means that several sub-bullets will create a ladder of dashes. What does a dash mean to a financial person? It is a *minus* sign, which means a negative.

You don't want your audience to go *there*. Avoid the subtle, subconscious, negative message sent by the dash. You could change the default symbol to a dot, which is an improvement, but it still clutters the slide. Get rid of the Morse-code effect of dots or dashes completely by indenting the sub-bullets. The space creates the offset; the space makes less out of more.

Finally, do we really need all those sub-bullets? The answer is usually "No." Sub-bullets often add a layer of complexity and an extra burden for the audience without any offsetting benefit. Remember, you, the presenter, provide the body text. You'll have ample opportunity during your presentation to amplify the details that support the headline bullets.

Now we have a clean graphic, as you can see in Figure 7.3.

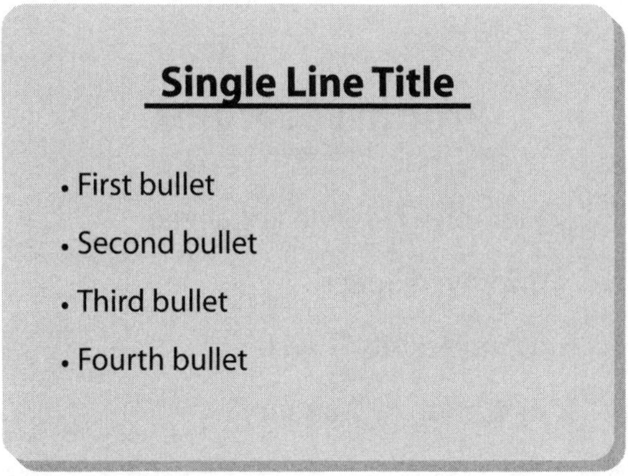

Single Line Title

- First bullet
- Second bullet
- Third bullet
- Fourth bullet

Figure 7.3 *The Less Is More bullet slide.*

To summarize:

- The Less Is More bullet slide contains one concept, expressed in a one-line title.
- The subtitle is best omitted.
- Bullets contain key words only, such as nouns, verbs, and modifiers. Avoid using articles, conjunctions, and prepositions. (Especially avoid using prepositions; not only do they add an extra word, they juxtapose and separate important words. Instead of writing a bullet such as "Strengths of Our Company," rewrite it as "Our Company Strengths.")

To make your bullet slide clean and crisp, try to follow the four-by-four formula: four lines down, four words across. Or, if the subject warrants, you can go up to six-by-four: six lines down, but still only four words across. A final benefit: With only one set of bullet symbols, visible at first glance, both you and your audience get a quick snapshot of the total idea.

Minimize Eye Sweeps with Parallelism

When all your bullets are parallel or similar in meaning (for example, a list of products, product features, or product benefits), the relationships will immediately be clear to your audience. However, whenever you create such a parallel list, pay particular attention to the grammatical form of the bullets. If you write each one in a different part of speech, you'll be forcing your audience to do extra work to grasp the logic. They'll have to reset their minds at the start of each bullet, as you do when you read the list shown in Figure 7.4.

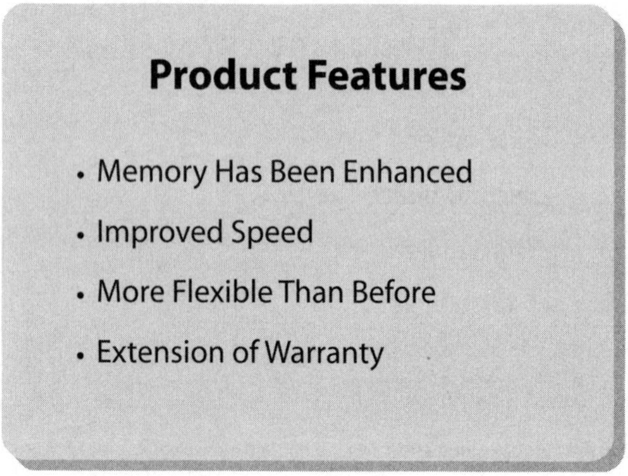

Product Features

- Memory Has Been Enhanced
- Improved Speed
- More Flexible Than Before
- Extension of Warranty

Figure 7.4 *Bullets in nonparallel form.*

Don't make me think! Each of these bullets represents a different grammatical form. The first bullet is a complete sentence with a passive verb, the second bullet is a noun modified by a preceding adjective, the third bullet is an adjective modified by a preceding adverb, and so on. Grammatical terminology notwithstanding, you can feel the inconsistency.

This problem is akin to the "floating decimal" dilemma in accounting, in which a series of figures is printed with irregular decimal places. This makes the numbers very difficult to read, anathema to any bookkeeper or accountant. In bullet slides, nonparallel construction of bullets is anathema to your audience.

To solve this problem, write any set of bullets in a list in parallel (grammatically the same) form, so that the similarity and relationships of the underlying concepts are obvious, as shown in Figure 7.5.

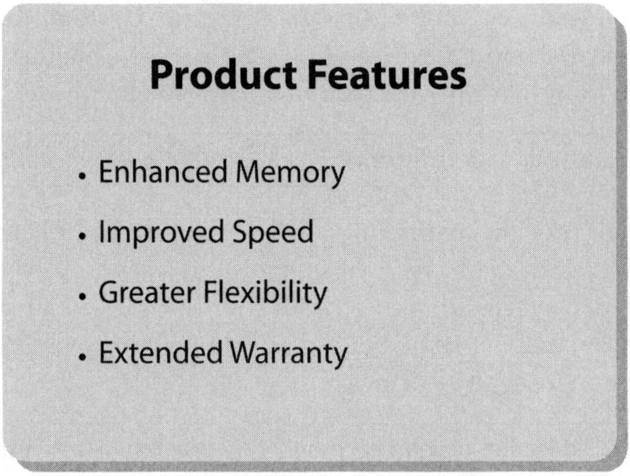

Product Features

- Enhanced Memory
- Improved Speed
- Greater Flexibility
- Extended Warranty

Figure 7.5 *The same concepts as in Figure 7.4, but in grammatically parallel form.*

In Figure 7.5, each of the bullets is in the same grammatical form: adjective plus noun. This parallelism in both meaning and form makes it easy for your audience to see the relationships. With this organization, you can display all four bullets at once. Your audience can absorb them quickly and then turn their attention back to you to listen as you add value.

Using the Build

Sometimes parallelism is not possible, so the relationships among the bullets are not readily apparent. And sometimes a single concept requires more than four bullets . . . perhaps as many as five or six or even eight bullets to fully explain one main concept. But showing a set of diverse bullets all at once is too much input for your audience's eyes and minds. It overwhelms their reception.

The effective solution for such cases is to build the list one bullet at a time. The build is easy to do with the Custom Animation feature in Microsoft PowerPoint or any other presentation graphics program. Reveal the first line, and explain and discuss it. Then click to reveal the next line, and explain and discuss it, until you reveal all the bullets.

When building your bullets, keep in mind Perception Psychology: Audiences find left-to-right movement natural and easy. Make it easy for your audience. Build your bullets by bringing them in from left to right. In Microsoft

PowerPoint, you can accomplish this by selecting Custom Animation, then Entrance, then Wipe, and then From Left. (Please see Chapter 12, "Animating Your Graphics," for a fuller discussion of animation techniques.)

Build your bullets by bringing them in from left to right.

As you build, provide *continuity* in your narration by linking one bullet to another as you click through. By *controlling* the revelation of your bullets, you keep your audience in synchronization with your flow. They cannot get ahead of you. Best of all, you can *add value* beyond each bullet by discussing, interpreting, and providing supporting evidence.

You can add value *beyond each bullet by discussing, interpreting, and providing supporting evidence.*

Bullet Levels

Some presenters are not satisfied with one level of bullets, so they use sub-bullets, thinking this will enable them to elaborate on their ideas. Sometimes two levels don't seem to be enough, so they use sub-sub-bullets. Occasionally, presenters completely surrender to temptation, and they go deeper, and deeper, and deeper, and deeper . . . all the way down into the right-brain basement (see Figure 7.6). The trouble is that most audiences can't follow the presenter's thought patterns that far down.

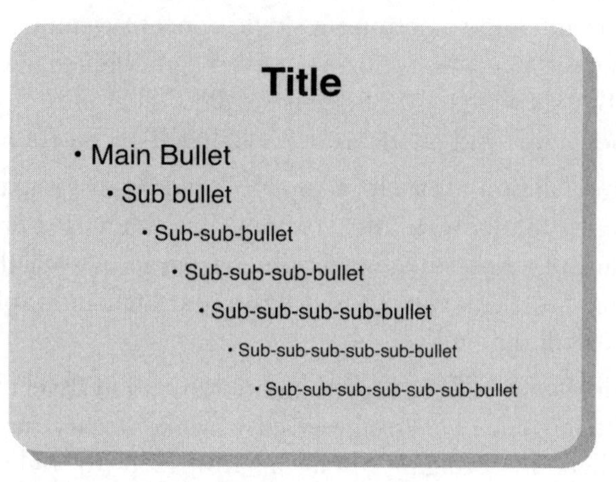

Figure 7.6 *Multiple levels of sub-bullets. How deep can you go?*

Do your audience a favor: Restrict yourself to only *one sub-bullet level*. A single level of sub-bullets (or no sub-bullets at all) keeps slides clear, crisp, and easy to read. It avoids forcing the audience into mental contortions as they struggle to track the presenter's internal logic. Additionally, avoid the Morse-code effect by eliminating symbols for sub-bullets. Use only an indent to create an offset. Less Is More.

Avoid the Morse-code effect by using no *symbols for sub-bullets. Use only an indent to create an offset.*

If you do use sub-bullets, put the same number of sub-bullets under each bullet, as shown in Figure 7.7. If the first bullet has two sub-bullets, then every other bullet should have two sub-bullets. The resulting symmetry creates a balanced image, as well as the message that your ideas are logical. The visual organization creates the subliminal message of *Effective Management*.

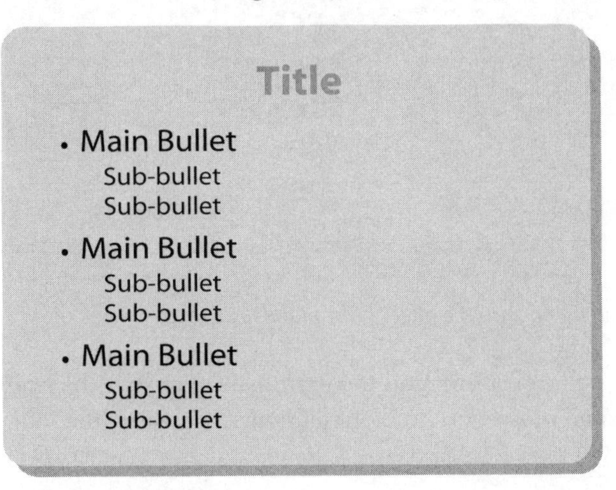

Figure 7.7 *Only one level of sub-bullets.*

Verbal Style

Now for some stylistic techniques that apply to all text slides, whether they contain bullets or sentences. These are seemingly minor points that your audience may never *consciously* notice, but have a definite subconscious impact on them and on how they perceive your slide, your message, and therefore you. Follow these simple and straightforward style guidelines whenever you create text slides.

Use Possessives/Plurals Correctly and Sparingly . . . if at All

This is a grammatical and stylistic point that surprisingly few people understand. The use of an apostrophe plus "s" ('s) to mark the plural form is simply bad English. An apostrophe plus "s" should be used only for contractions (words from which one or more letters have been dropped, such as I'll, can't, you'd, and he's) and possessives (such as IBM's new chairman, the company's headquarters).

Nonetheless, many people mistakenly use an apostrophe plus "s" in plurals, especially when pluralizing acronyms or numbers, as shown in Figure 7.8.

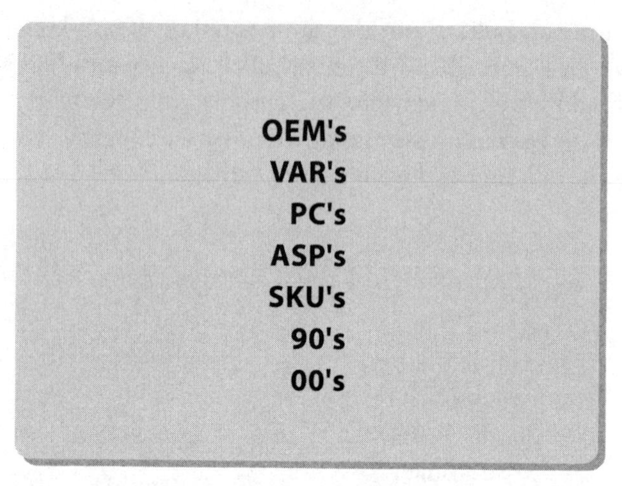

Figure 7.8 *Using an apostrophe plus "s" for plurals is wrong.*

This approach can compound the grammatical mistake by leading to momentary confusion and more work for your audience. Consider the following sentence:

> DVD's produce sharper images than VHS's because a DVD's resolution of 500 lines is greater than a VHS's resolution of 240 lines.

The first DVD and VHS references are plural, so they don't need an apostrophe. The second references are possessive, so their apostrophes are correct. The incorrect apostrophes will confuse the reader.

Using acronyms is risky, because they may be unfamiliar to part of your audience. If you *do* use an acronym, the correct way to turn it into a plural form is with a lowercase "s" and *no* apostrophe. The same applies to a number or to any other word that needs to be expressed as a plural, as shown in Figure 7.9.

If you do *use an acronym, the correct way to turn it into a plural form is with a lowercase "s" and* no *apostrophe.*

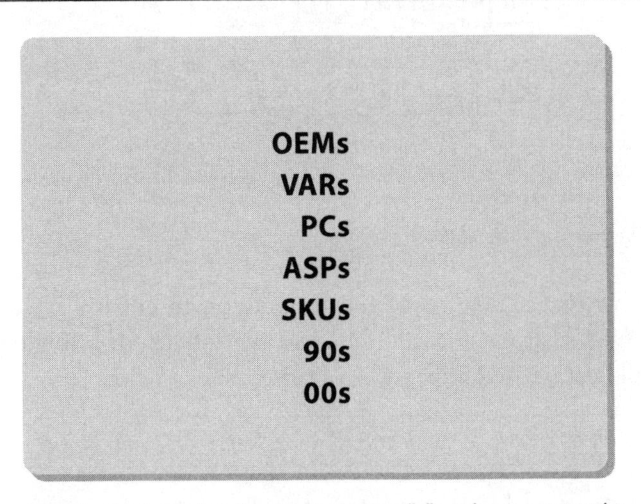

Figure 7.9 *The correct way to make plurals: lowercase "s" and no apostrophe.*

Actually, you can and should avoid using an apostrophe plus "s" *at all* in your presentations. Although an apostrophe plus "s" is grammatically correct for possessives, you can exercise literary license in the interest of Less Is More. Eliminating an apostrophe plus "s" eliminates extra characters and makes the text easier to read and comprehend. Consider the difference between "IBM's New Chairman" and "New IBM Chairman."

Keep Your Font Choices Simple

Most graphics programs, including Microsoft's PowerPoint, provide dozens of different type styles, known as *fonts*. Some cost-conscious presenters seem to think, "We've paid for all these fonts, so we should use all of them." The result is a slide that looks like a ransom note, as shown in Figure 7.10.

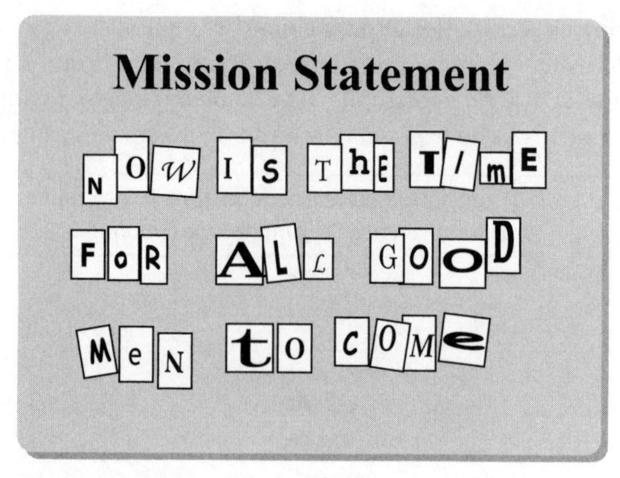

Figure 7.10 *The ransom note slide.*

For all your text slides, resist the temptation to get too creative with font choices. Use two or, at most, three different type styles for a single presentation. The result will be a unified look and feel that conveys a clear, consistent message.

Use two or, at most, three different type styles for a single presentation.

One easy way to add creative styling and remain simple is to use one font style for titles and another for bullets. Another is to use two sizes of the same font, a larger one for titles and a smaller one for bullets. You can add further styling by choosing one color for titles and another for bullets, or you can make one of your choices an italic version of the same font. Less Is More is particularly applicable when it comes to typographic choices.

Proportional Spacing

Let's say you've followed the four-by-four rule for your bullet slide, and it looks like Figure 7.11.

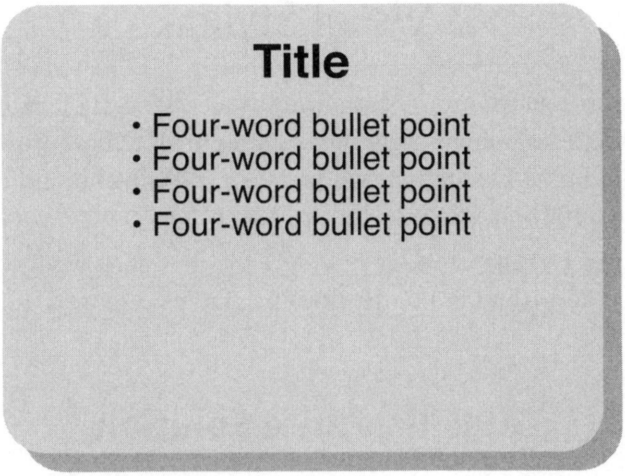

Figure 7.11 *Four by four . . . but are there more bullets?*

See the problem? All the bullets are bunched up in the top half of the slide, leaving the bottom half empty. This imbalanced design has the secondary negative effect of setting up an anticipation that will not be resolved. *Are there more bullets?* You can easily remedy this problem with proportional spacing: Distribute the bullets evenly over the entire slide, as shown in Figure 7.12.

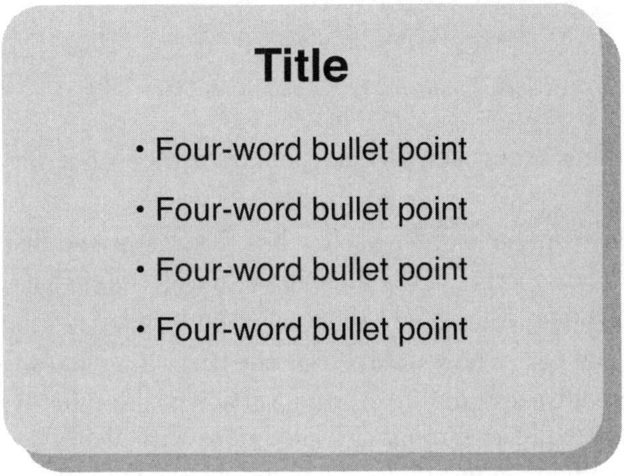

Figure 7.12 *Proportional spacing.*

Visual Style

Many business presentations are supported by a collection of slides that are only text . . . no pictorial, relational, or numeric slides at all. Not only does this border on the Presentation-as-Document Syndrome, it looks bland and boring. Twenty all-text slides in a row can produce the MEGO effect.

Furthermore, reading an all-text slide feels like hard work, even when the slide is designed according to the previous principles. Figure 7.13 is an example.

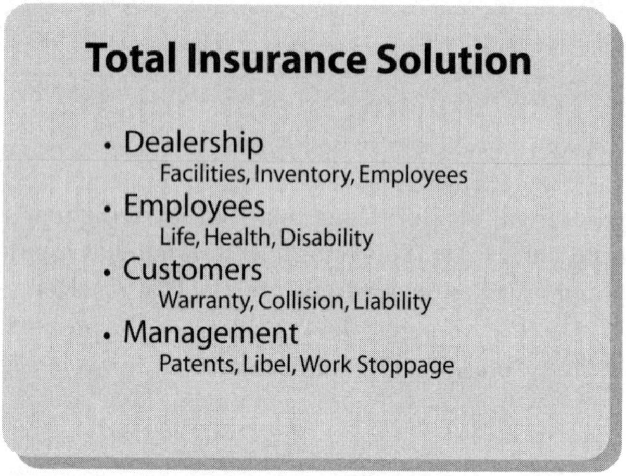

Total Insurance Solution

- Dealership
  Facilities, Inventory, Employees
- Employees
  Life, Health, Disability
- Customers
  Warranty, Collision, Liability
- Management
  Patents, Libel, Work Stoppage

Figure 7.13 *An all-text slide, multiplied by 20, equals MEGO.*

There's nothing wrong with this slide, but it's certainly not interesting or even visually appealing.

You can add styling to an all-text slide like this without adding or subtracting a single letter. The secret is to use the simple design tools that are included in Microsoft PowerPoint. Figure 7.14 shows an example.

By embedding text in boxes, and giving the boxes lines and shadows, you create a much more attractive and interesting version of the slide. It also makes the information easier to read. Grouping related items visually reduces the interpretive work your audience must do to follow the flow of your ideas.

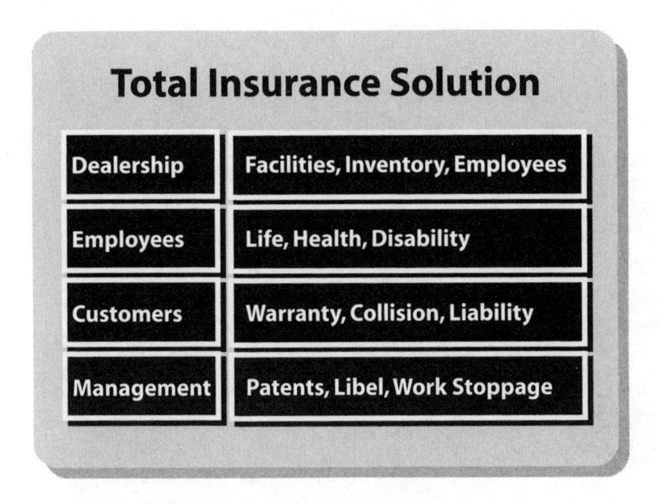

Figure 7.14 *The same text as Figure 7.13, but with styling.*

Microsoft PowerPoint provides a wealth of design options for adding styling to your text slides with a vast array of shapes, colors, elements, and formats. The Office 2007 version of PowerPoint has a new feature called SmartArt graphics in which many different style effects can be applied simultaneously with a single mouse click.

You can also create emphasis by using a manual technique called *reverse out*, which sets off a headline from the rest of your text by reversing the background and font colors, as shown in Figure 7.15.

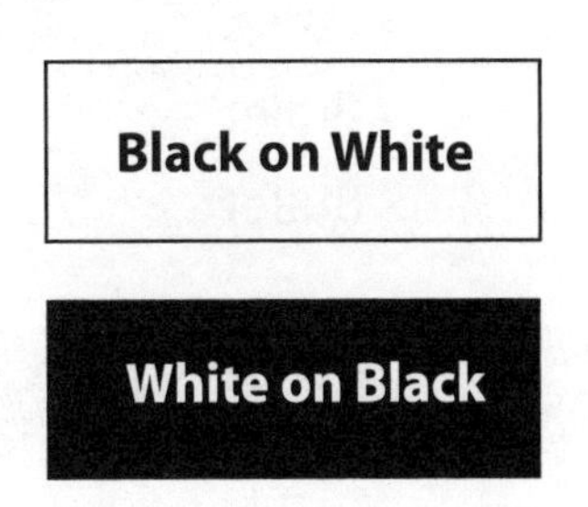

Figure 7.15 *The reverse-out effect.*

You can also enhance the design of the background by adding stripes, borders, edges, and cornices. There are so many interesting design options that it would be easy to get carried away. For example, one option is gradient shading, as shown in Figure 7.16.

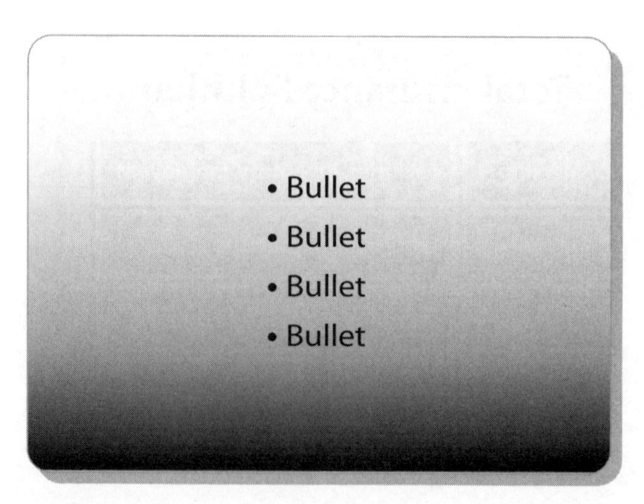

Figure 7.16 *Gradient shading.*

This looks cool at first. But then you realize that it's difficult to read the last bullet. Not to worry; there's always another bell or whistle in the graphics toolbox. You can add a double gradient, as shown in Figure 7.17.

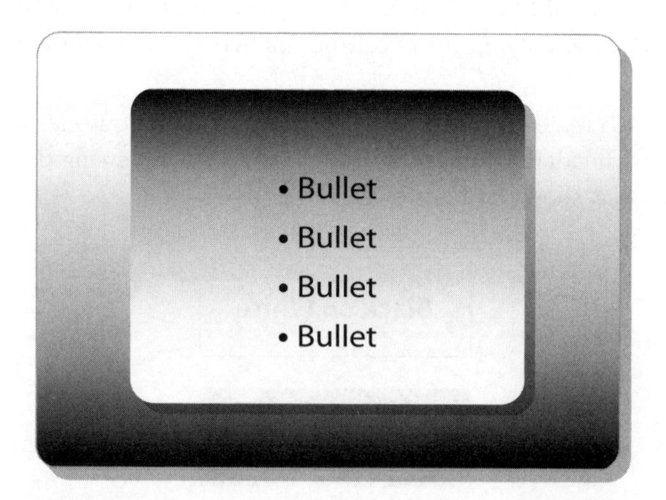

Figure 7.17 *Double gradient shading.*

But now you realize that you can't read the first bullet!

The point here is the same as with font choices: Remember the Less Is More principle. Many presenters get carried away with the plethora of fancy graphic toys available to them. Choose one or two graphics effects that will enhance the clarity and attractiveness of your slide design, and use *only* these throughout your presentation. Select a limited color palette, two or three colors at most, to complement the colors in your company logo, and use them consistently. For example, if you use a gold border and a royal blue headline on your first slide, then every slide should contain a gold border and a royal blue headline.

Choose one or two graphics effects that will enhance the clarity and attractiveness of your slide design, and use only *these throughout your presentation.*

Finally, optimize all your text slides by observing the following simple guidelines.

Text Guidelines

- Create a consistent look and feel, and maintain it throughout.
- Be consistent in your choice of font, as well as in your choice of case.
- Keep font size to a minimum of 24 or 28 points.
- Avoid abbreviations at all costs.
- Add shadows and bolding to make all text more legible.
- Use sharp contrast: light text on a dark background, or vice versa.
- Insert your company logo, but don't make it look like a neon sign; treat it instead as a watermark, with a subtle, embossed effect.
- Avoid the clutter caused by recurring slogans, datelines, copyrights, and the ubiquitous "Company Confidential" warning in the periphery of every single slide.
- Use blank space. You don't have to fill every nook and cranny of every slide with information. Costly newspaper advertisements often use valuable white space to set off text. Look at them, see the difference, and follow their example.

You'll notice that I haven't mentioned the differences between serif and sans serif fonts, or text justification right, left, or center. These typographic fine points are matters of individual taste, and as the Latin proverb tells us, *De gustibus non est disputandum* (There's no arguing taste).

Follow the preceding guidelines, and your text graphics will be simple, consistent, and logical, reinforcing the subliminal message of *Effective Management*.

In the next chapter, you'll see how the same basic principles that create winning text graphics also apply to numeric graphics.

Chapter Eight

Making the Numbers Sing

The Power of Numeric Graphics

Numbers play a key role in any business presentation. Revenues, units shipped, profits, and market share are the hits, runs, and errors of the business scorecard, and everyone in business understands their importance.

However, not everyone in business feels equally at home when it comes to numbers. There are the green-eyeshade types who immediately recognize key trends and who can quickly pick out the most important item in a column of figures. Then there are the rest of us, who need a little time and a lot of context to fully grasp the meaning of a profit-and-loss statement or balance sheet.

In any presentation, you want both the number-fluent and the number-challenged audience members to understand and agree with you. Skillfully designed numeric graphics can help achieve that. They translate digits and decimals into visual images that make abstract relationships concrete and much easier to recognize.

Unfortunately, many of the numeric graphics used in presentations serve to obscure rather than clarify the facts. All too often, the graphics are sheer Data Dumps, loaded with needless information, poorly organized, and visually cluttered. Such numeric slides take a long time to explain and even longer for the audience to understand. Often, the audience members decide that such slides aren't worth the effort, and they give up. This is how many important presentations get derailed.

This needn't happen with your presentations. The same basic principles you learned in Chapter 6, "Communicating Visually," Presenter Focus, Less Is More, and Minimize Eye Sweeps, apply to numeric graphics. Let them be your guidelines to creating clear and effective images that will drive your story home and reinforce your key ideas.

Bar Charts

Let's begin with the Problem/Solution approach by looking at an all-too-typical numeric slide, the bar chart shown in Figure 8.1. You've probably encountered this type of graphic in many presentations. This example depicts six years of steady sales growth in the history of an up-and-coming company, information that should play an exciting role in telling the company's story.

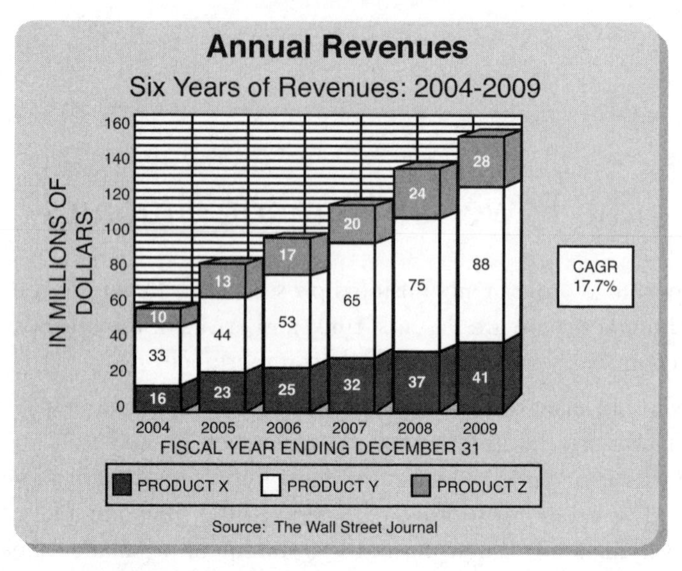

Figure 8.1 *How not to design a bar chart.*

There's plenty of information here. In fact, the problem is that there's *too much* information for a presentation. If this chart were in a document, such as a business plan or annual report, the reader, as the audience to the absent writer, would need all this data to understand the chart and identify the figures. The reader would also have close-up access to the document. But in a presentation, the audience is forced to make multiple *involuntary* eye sweeps across the distant enlarged image to *see* all the data, while simultaneously their ears are listening to the presenter and their minds are *processing* all the data. All this ratcheting around and thinking is excessive sensory activity that disconnects the audience from the presenter. The slide becomes the focus of the presentation.

Sound familiar? These same underlying problems have a negative impact on the graphics, just as they do on the story, drastically reducing the potential to capture your audience.

Here's how to communicate the same information successfully: First, notice how much print clutters the slide. The title and subtitle essentially repeat the same information. There are two labels, one at the left side of the slide ("IN MILLIONS OF DOLLARS") that is standing on its end, and one sprawled along the bottom of the slide ("FISCAL YEAR ENDING DECEMBER 31"). Each label takes up a great deal of space while providing information that is either very simple or not very important. There are more than 30 numbers on the slide: nine along the left scale, 18 superimposed on the six bars, and the six dates along the bottom edge of the graph. The CAGR (Compound Annual Growth Rate) sticks out to the right like an outrigger. The source credit to *The Wall Street Journal* is in a font size so tiny it rings of the fine print in a shady contract. The six bars depicting revenues are each subdivided into three different colored sections. And all of this data is backed by grid lines that resemble bargain-basement, bamboo-slat Venetian blinds.

Clearly the person who designed this slide has never heard of Less Is More.

Notice how much work your eyes (and your mind) must do to absorb all this information. It's not only a matter of wading through all the words and numbers to decipher which ones are important and which aren't; it's also a matter of attempting to draw connections among the disjointed parts of the graph. In short, this slide is a visual mess. How can we improve it?

We can start by simplifying and cleaning up the unnecessary verbiage. Since the title and subtitle are redundant, we can eliminate the subtitle. The graph itself clearly contains six bars, which are labeled by year, so it isn't necessary for the subtitle to spell out the number of years covered.

Rather than devoting so much space to the labels "IN MILLIONS OF DOLLARS" and "FISCAL YEAR ENDING DECEMBER 31," we can use simple abbreviations. Nor do we need to spell out full dates. In this context, if a bar is labeled "'04," everyone in the audience will understand that it represents 2004 rather than 1004 or 3004. Give *The Wall Street Journal* its due by making the font size legible. Finally, does your audience really need to know the exact dollar figure for each product sold? Probably not; it depends on the point of the slide. If the purpose is to demonstrate total revenue growth, then the audience can see the relative amounts of each product by the colors. Remove the numbers in the bars.

This is a distinct improvement. We've substantially reduced the amount of work the audience must do to understand the graph. But there's still more to be done. To identify the different colored parts of the bars, the audience members must shift their eyes up and down repeatedly to the legend at the bottom, like a bouncing yo-yo.

Only when the audience discovers the legend at the bottom of the slide, which matches the colors to Product X, Product Y, and Product Z, does the explanation become clear. You can feel the movement here on the page, where the distance is only several inches. Imagine the feeling when you traverse a space of several feet on a projection screen. *Minimize Eye Sweeps.*

Furthermore, anyone in the audience who wants to figure out the value of a particular bar must move his or her eyes back and forth, left and right, several times between the scale and the bar, like a fast and furious ping-pong match. A futile match, too, because, at a distance, the human eye can't find the exact tick mark. Would any businessperson respond favorably to a proposal with inexact figures that could be viewed a few million dollars in either direction? Hardly.

We can make some simple adjustments to clear up these problems . . . again, by applying the *Minimize Eye Sweeps* rule.

Look at Figure 8.2.

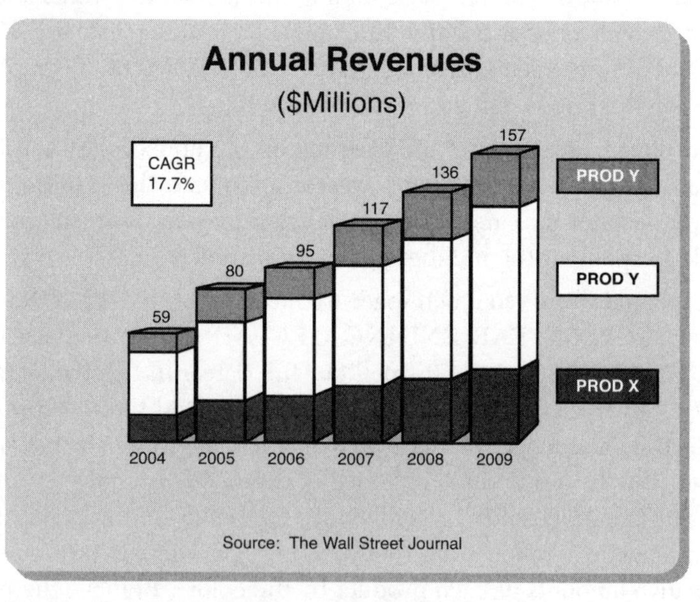

Figure 8.2 *The same bar chart, simplified and clarified.*

By removing the scale from the left and placing the revenue totals directly above the bars, we see the growth trend in one sweep of the eyes. By removing the legend at the bottom of the graph (along with the usual trifling little squares) and simply labeling the three different colored bar segments in matching colored text boxes along the right margin, the eyes identify products X, Y, and Z at the end of the eye sweep. Simpler still would be to eliminate the stacks and create separate charts for each product.

Notice the three-dimensional effect on the bars. Some people like this effect; others hate it. As with typography in the previous chapter, these are matters of individual taste, and the same Latin proverb applies: *De gustibus non est disputandum.* There's no arguing taste.

The simplified version of the slide will have much more impact in your presentation. The important story it tells about your company's impressive sales growth will hit home much more forcefully. *Any* numeric slide can be dramatically improved by eliminating unnecessary words, numbers, scales, and legends.

Any *numeric slide can be dramatically improved by eliminating unnecessary words, numbers, scales, and legends.*

Pie Charts

Figure 8.3 shows a typical pie chart (also called a circle chart). This kind of chart is useful for showing a total amount divided into subordinate parts; in this case, the chart shows how a company's sales are divided geographically. At a glance, it's easy to see the relative share of the whole that each part, or each wedge of pie, represents.

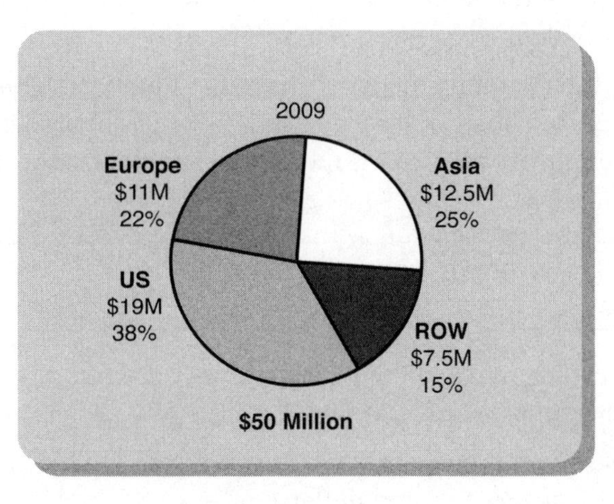

Figure 8.3 *A typical pie chart.*

Unfortunately, the chart is needlessly cluttered and confusing to read. Stacking the name of the sales region, the sales figure, and the percentage (such as "Europe," "$11M," and "22%") forces the viewer to pause and sort out what each element means.

Now look at the version shown in Figure 8.4.

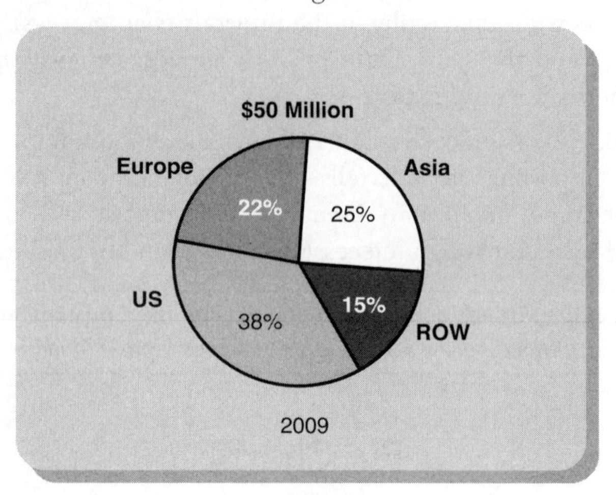

Figure 8.4 *A simplified pie chart.*

The geographic labels remain outside the pie, but the percentages are now inside the wedges. This arrangement is clearer because text usually takes up more space than numbers. If any wedge is too small to contain its number, use a callout; that is, place the number just outside the wedge, with a narrow line to indicate where it belongs.

Notice that by separating the labels from the numbers, both are much easier to read. Notice, too, that we have omitted the dollar figures shown in the first version of the graph. In a pie chart, the relative size of each wedge is the most important information.

In a pie chart, the relative size of each wedge is the most important information.

Finally, note that we've shifted the date to the bottom of the pie. That's where the timeline appears in most business charts.

Follow these general guidelines when you create a pie chart, and it will be easy for your audience to read and understand.

Typography in Numeric Graphics

The label on the left side of Figure 8.5 is "stacked." Type set this way is very hard to read. Think about it: The stacked label, containing eight letters, causes your eyes to make seven carriage returns. Ouch!

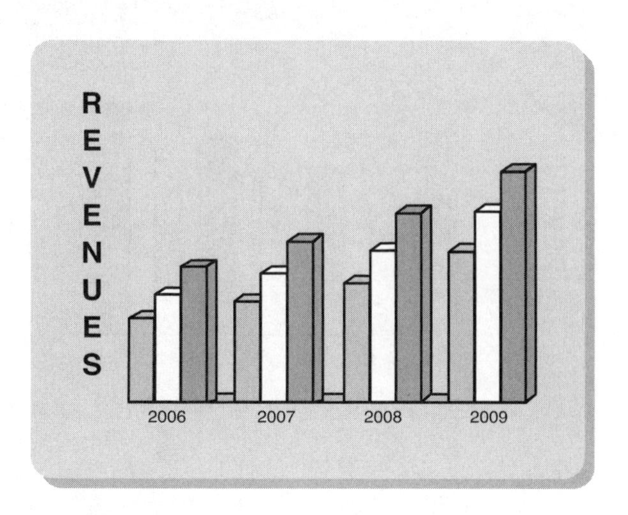

Figure 8.5 *The label on the left is "stacked."*

This problem is easy to solve, as shown in Figure 8.6.

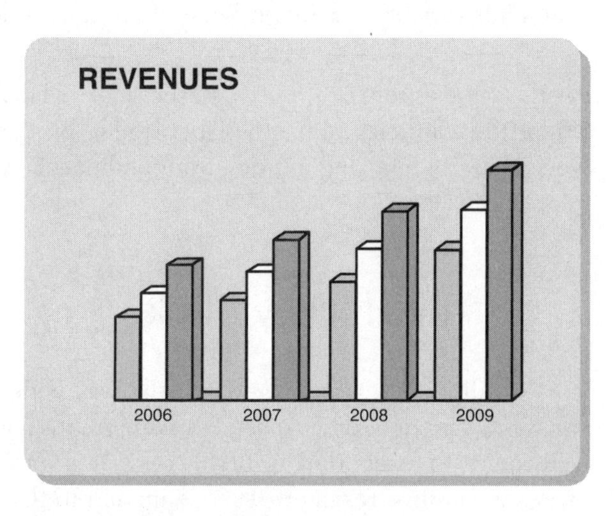

Figure 8.6 *The more legible horizontal label.*

Use a horizontal label like the one shown in Figure 8.6, which makes the chart much clearer and easier to read.

A close cousin of the stacked vertical label is the vertical label on end, shown in Figure 8.7.

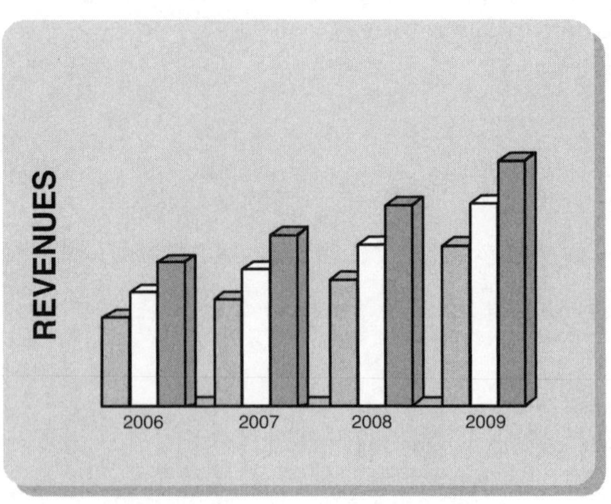

Figure 8.7 *The vertical label on end.*

This practice is a carryover from documents. In a document, the reader can rotate the document to read the word "REVENUES." In a projected presentation slide, however, the audience members are forced to rotate their heads. This often seems like an adult variation of Simon Says: "Left ears on left shoulders, place!"

The solution is the same as for the problem of the stacked label (please refer back to Figure 8.6 for the solution). Make the label legible by aligning it on the horizontal axis. Now you're minimizing not only your audience's eye sweeps, but their head movements as well.

The Hockey Stick

Another example of conditioned eye movement is the way audiences perceive and react to information. Through long experience with graphs and charts, businesspeople are accustomed to responding favorably to what is known as the *hockey stick*, which expresses positive results that move up and to the right. A trend sloping in the downward direction implies negative results and is therefore a counterintuitive movement.

If you were creating a graph to compare your company's results (sales, profits, product performance, or customers) against those of your competition, would you show it using bars like those in Figure 8.8?

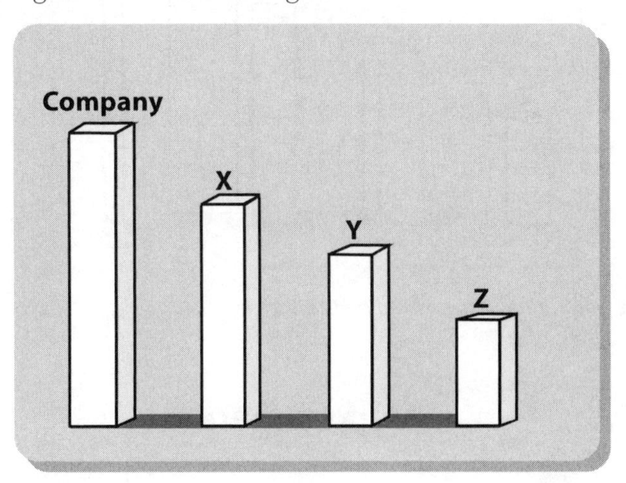

Figure 8.8 *The reverse hockey stick, going down.*

I hope not. The results are heading downhill. Instead, arrange the bars as shown in Figure 8.9, up and to the right. The hockey stick now expresses rising action. Furthermore, the bars end with your company name. By placing your company name at the final, climactic point, not only do you tacitly take a superior position, but your company name becomes the last word your audience will see and remember.

Unfortunately, many presenters, as well as many professional graphic artists, design their bar charts counter to these cultural implications. They are under the mistaken assumption that the most important figure should come first. While this reasoning seems logical, it is contradictory to the more powerful psychological forces. Figure 8.10 is a reproduction of an actual newspaper advertisement, doctored slightly to protect the guilty (we've used the fictitious bank name "Comet").

Comet wanted to brag about its leading market share over all its competitors in corporate debt underwriting. Although Comet was proud of its achievement, its advertising designers set up the bars with movement *downward* rather than upward, counterintuitive to positive results.

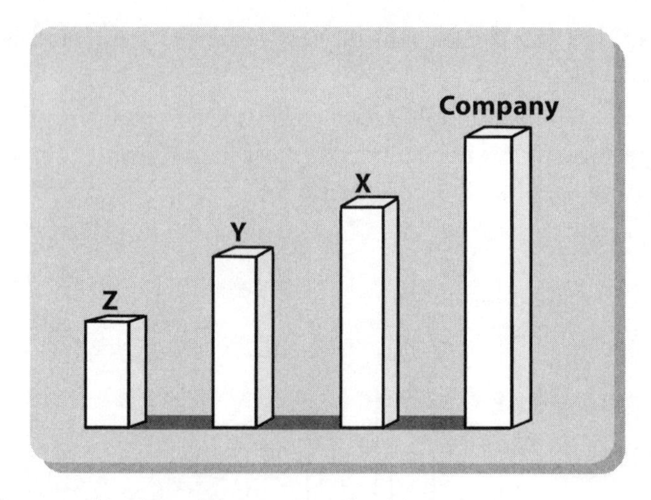

Figure 8.9 *The natural hockey stick, up and to the right.*

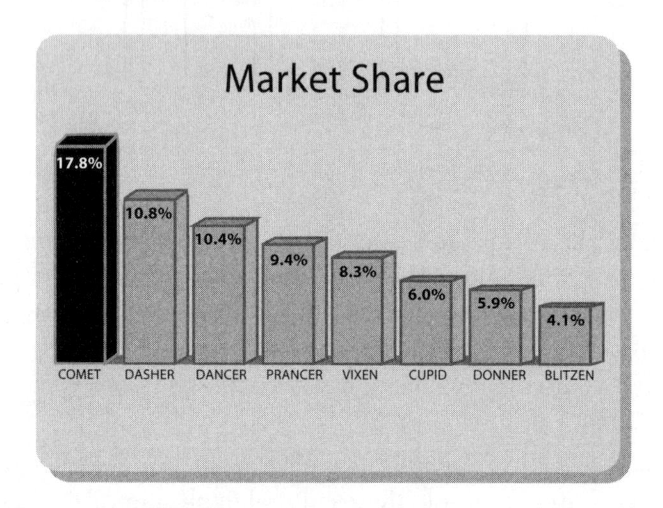

Figure 8.10 *A violation of the hockey stick rule.*

Figure 8.11 shows how Comet's ad should have been depicted.

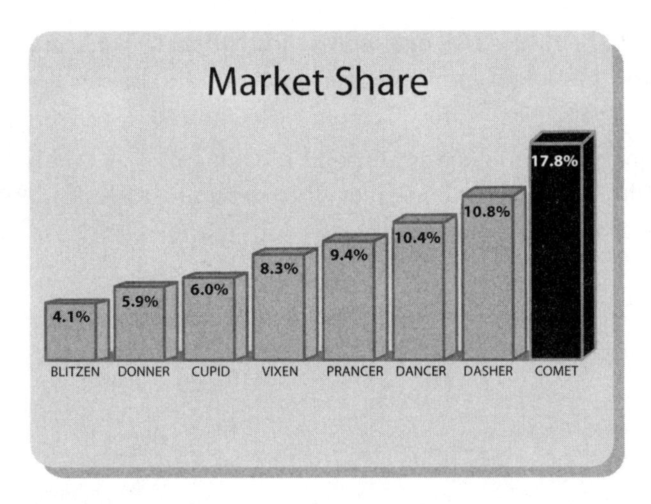

Figure 8.11 *Follow the rising action.*

Subtle? Yes, But . . .

The stylistic points presented in this chapter may strike you as picky. "Oh, come on," you might say. "The members of my audience aren't graphic artists or designers. Will anyone *really* notice if I arrange the bars in the graph in reverse order or stack the letters in the vertical label?"

Maybe not . . . maybe not consciously. But our cultural conditioning to react to visual cues is very deep and powerful . . . so much so that, even when your audiences don't consciously recognize such design flaws, they respond to them with a sense of unease, uncertainty, or dislike. They may not realize what's bothering them; at most, they might think, "Those slides look a little odd" or "There's something here that's not quite right."

Is this a serious problem? It can be. Remember, it is likely that that your efforts to persuade your audience will be competing against powerful opposing forces, ranging from your direct competitors to the subtler forces of indifference, apathy, and inattention. It's easy for anyone to doubt your expertise, to question your motives, to be distracted, or simply to lose interest. You can't afford to overlook any factor that might influence your audience, no matter how subtle it may appear.

Persuading your audience to respond to your call to action is almost always an uphill battle. Why make it harder, even 10 percent harder, by designing graphics that work against your message? Make your graphics work for you.

Chapter Nine

Using Graphics to Help Your Story Flow

Company examples:

- **Modex Therapeutics**
- **Intel**

The 35,000-Foot Overview

In the past two chapters, we focused on designing slides that convey information clearly and effectively. You've drilled deep into many details about everything from font styles to graph labels. Now it's time to take a step back and return to one of the earlier concepts in this book: specifically, to *flow*.

Every communications medium has its own techniques for helping its audience remain oriented and follow the flow. Think about text, where the reader is the audience to the writer. In a book, a magazine, a newspaper, or a printed report, the designers and editors provide the reader with many tools to help track the writer's flow: the table of contents, the index, and the running heads along the top or bottom of the page. Even more important, the reader has the luxury of random access to the writer's material. The reader can *navigate* through the text independently by visiting and revisiting the table of contents or the index, and by flipping backward and forward as often as needed to follow the overall flow. Think of Russian novels, where all those multiple names and nicknames of the characters are listed up front. Think of plays where the *Dramatis Personae*, or cast of characters, precedes the first scene. Through long practice, readers are accustomed to steering through the structure of written texts on their own.

Presentations are different. In a presentation, the audience can access the presenter's material only in a *linear* fashion: one slide at a time, one spoken sentence at a time, and all this under the presenter's control. Once a slide disappears from the screen, it's gone forever. The audience doesn't have the opportunity to flip back and forth at will to clarify the presentation's flow.

Just as in the telling of the presentation story, the audience receives the visual content at the level of the trees. And, just as in the telling of the story, the presenter's job is to rise up from the trees and give the audience a view of the entire forest. As always, the presenter's job is to *navigate* the audience's minds, but the presenter can also navigate the audience's eyes.

The Flow Structures in Chapter 4 and *Telling 'em what you're gonna tell 'em* in Chapter 5 can help your audience follow your flow, but both of these are purely verbal techniques. Graphics can help express flow, too. You can design your slides to convey the connections among your ideas. You can use a set of simple visual tools to create continuity and to help your audience track the overall logic of your presentation.

Before introducing these tools, however, it's important to step back and check the flow of your presentation, to examine it as a whole, by taking a 35,000-foot overview. One valuable tool to do this is the Storyboard Form, illustrated in Figure 9.1.

Television and film directors use storyboards to plan their end products, whether it is a 60-second commercial or a multimillion-dollar special-effects epic. They map out the camera angles of each scene and then envision how the individual scenes will combine into a whole sequence. We provide our Power Presentations clients with a paper version of the storyboard as part of our programs. (You can download a copy of this form by visiting our website, www.powerltd.com.) The Microsoft PowerPoint version of this aspect is called Slide Sorter view. Both options provide a panoramic view of your story.

This view lets you see all the slides for a given presentation at a glance, a perspective that minimizes your focus on details and offers a broader outlook of the landscape. It's an efficient planning tool that helps you check the progression of your story. Note that there is also a small rectangular box beneath each slide image where you can enter notes on your narrative, further clarifying your flow.

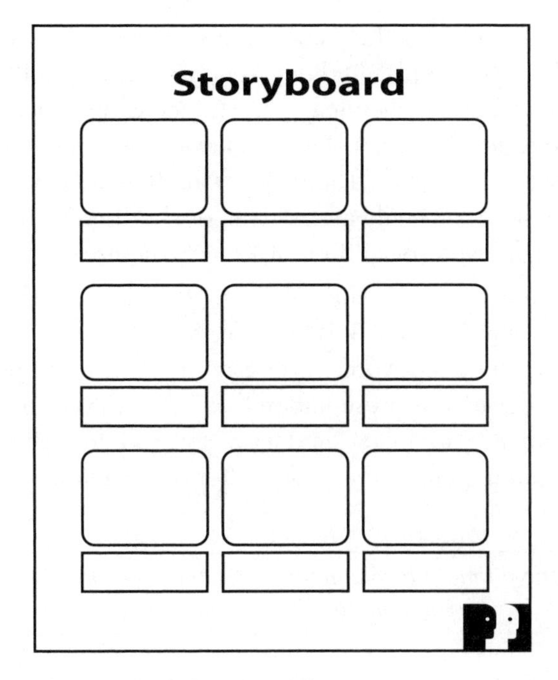

Figure 9.1 *The Power Presentations Storyboard Form.*

Figure 9.2 illustrates how the panoramic view works. Look at your slides in the Storyboard Form. See them in groups, reflecting the Flow Structure you chose for your presentation. In this example, the presentation includes four parts: an Introduction, an Opportunity section, a Leverage section, and the Conclusion. Each section is supported by a group of slides, and within each section, every slide should fit logically into place.

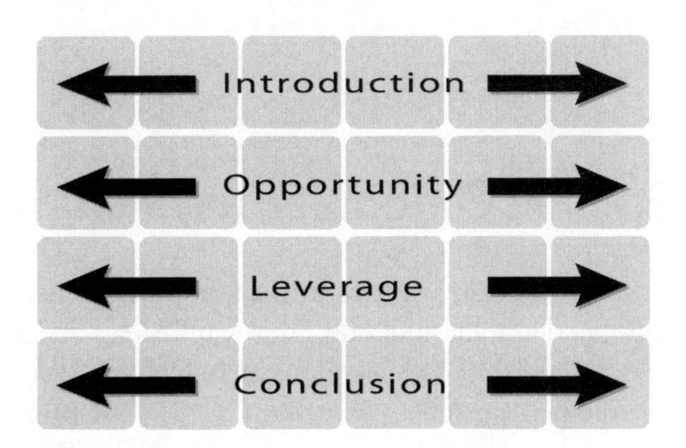

Figure 9.2 *The panoramic view.*

When you scan slides in this view, it should become evident when some slides don't fit where you've placed them. You may want to shift them to a different part of the sequence; or they may be extraneous, in which case, you might want to eliminate them altogether. The connection between any one part of the presentation and the next should also be clear. If it isn't, you might want to rethink your sequence. You could add, delete, or shuffle slides so that the logic is evident; or you might even try a different Flow Structure.

The ultimate technique for checking your flow is to read only the titles of your slides, as shown in Figure 9.3. If you can trace the logic of your entire presentation by reading these few words, bypassing the bullets, graphs, or other content, you've created clarity. A presentation that achieves this inner logic makes it easy for your audience to follow and easy for you to deliver. In Microsoft PowerPoint, you can use either the Slide Sorter or the Outline view to read only the titles.

The ultimate technique for checking your flow is to read only the titles of your slides.

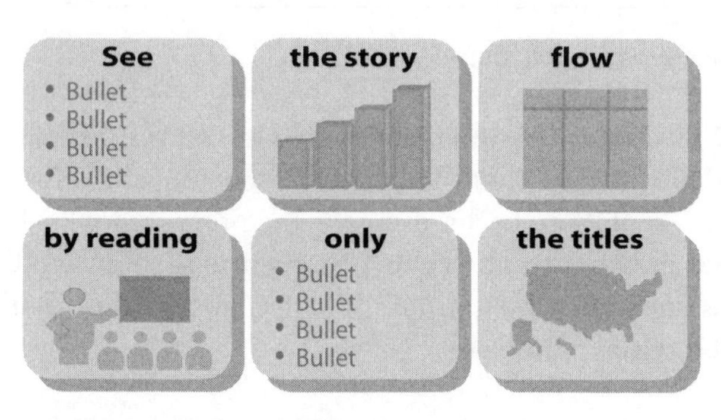

Figure 9.3 *The ultimate flow check.*

Graphic Continuity Techniques

Once your verbal logic is clear, you can turn to graphics to help communicate your flow. Well-designed graphics not only convey information clearly and attractively, they also help establish the connections among ideas. Again, the panoramic view offers an excellent perspective on the flow of your presentation. If the continuity of your graphics is not apparent, it might look like Figure 9.4.

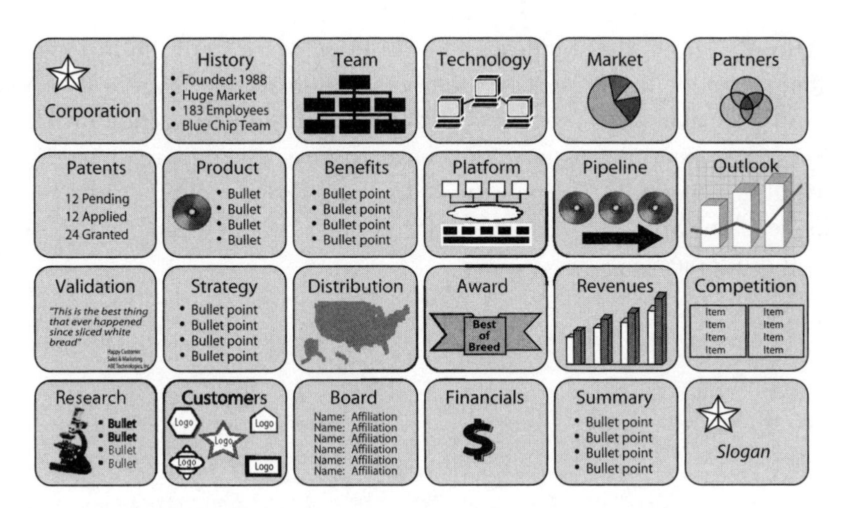

Figure 9.4 *No two slides alike.*

Try to scan this storyboard from slide to slide. Because no two slides look alike, the presentation begins anew with each slide, forcing you to adjust repeatedly to new content, new form, and a new look. Now think of your audience. Without the luxury of the panoramic view, they will see only one new slide *at a time*, and they will be forced to work *even harder* to adjust to each new content, new form, and new look. By the end of the presentation, they will be exhausted and, most likely, confused. *Don't make them think!*

There are five proven graphics techniques that you can use to help express the flow of your presentation in your slides. Let's look at each of these techniques, along with illustrative examples drawn from actual business presentations.

1. Bumper Slides

The first continuity technique, and the simplest, is the *Bumper slide*. This technique is drawn from the field of publishing. Look at any lengthy nonfiction book. You're likely to find that it is divided into chapters and parts. Most often, each part is separated from the next part by a full page containing only the part number and a title.

The purpose: closure of the outbound section, and a lead into the inbound section. The advantage: The reader's mind is cleared of the last subject and primed for the next. In a presentation, the Bumper slide signals a transition to a new section. Think of the Bumper slide as the sorbet that is served between courses of a fine meal to cleanse the palate.

Figure 9.5 shows the simplest and most effective form of the Bumper slide. It contains a *single* line of text (remember: Minimize Eye Sweeps by avoiding wordwrap). That one line previews the contents of the next section of the presentation. For example, you might have Bumper slides reading "Market Opportunity," "Unique Technology," "Industry Leadership," and "Business Results," each of which would serve to introduce a set of ensuing slides covering each of these topics.

Figure 9.5 *A Bumper slide with centered text.*

To differentiate the Bumper slide from all the other slides, center the text both horizontally (left to right) and vertically (top to bottom).

A simple image or symbol could also serve as a Bumper slide. It could be your company logo, an image relevant to the theme of your presentation, or an icon that suggests the content of the subsequent portion of the presentation.

You can also use your agenda as a Bumper slide (see Figure 9.6) and, with it, track the overall outline of your presentation. For this approach, each Bumper slide displays the complete agenda with the upcoming item *highlighted* . . . as opposed to dimmed, which causes your audience to squint.

The Bumper slide tells your audience exactly where they are in the presentation, as well as where they've been and where they're going. Think of this as a progressive agenda. A caution: Use this approach only with longer presentations . . . say, 30 minutes or more. In shorter presentations, if you reprise your agenda after just one or two slides, your audience will feel patronized. They might say to themselves, "I got it! You don't have to remind me!"

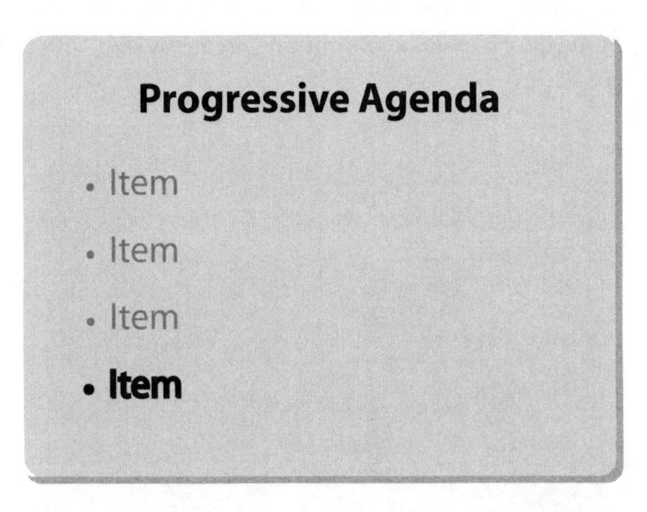

Figure 9.6 *The agenda as a Bumper slide.*

In longer and more complicated presentations, your audience will appreciate these milestones. Here's an example from the world of biotechnology:

In 1999, Dr. Jacques Essinger, the CEO of Modex Therapeutics, a Swiss biotech company focused on cell therapy, engaged me to help him and his team develop their IPO road show. The company went public the following year on Switzerland's equivalent of the NASDAQ, the Nouveau Marche (SWXNM), and it performed very well. In 2002, Modex merged with IsoTis and now operates under the name IsoTis Orthobiologics, listed on the Zurich, Amsterdam, and Toronto stock exchanges.

In mid-2001, the biotech sector in general, and growth companies like Modex in particular, fell out of favor with investors, and Modex's share price dropped. Jacques formulated a new strategy to acquire additional related technologies that could produce more revenues. When the acquisitions were done, Jacques called on me to help him prepare a new presentation to announce the new strategy to the Swiss financial community.

Since the focus of this chapter is graphic continuity, I'll discuss only the heart of Jacques' presentation, where he wanted to describe his newly expanded core business to his potential new investors, none of whom was a biotech scientist.

Jacques had a very crisp and strong opening sequence at the front end, and two brief sections on strategy and finance at the back end, but the core of his presentation is an exemplary illustration of the use of Bumper slides. He started by identifying Modex's original and continuing focus on skin with a simple circle, as shown in Figure 9.7.

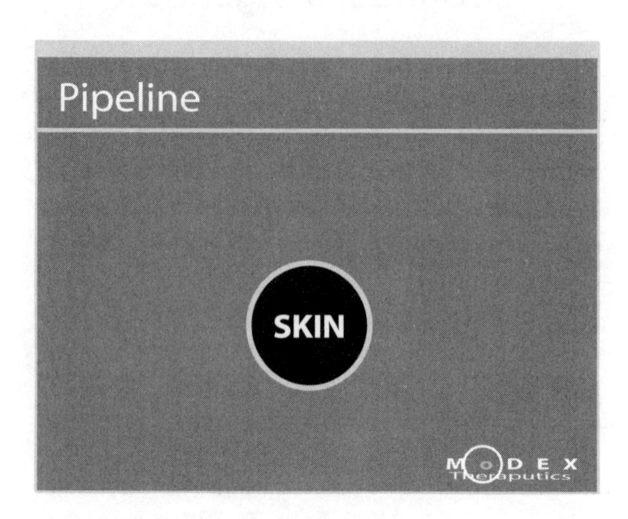

Figure 9.7 *Modex Therapeutics' business focus slide.*

Then, using the Microsoft PowerPoint animation feature, he sprouted three smaller circles from the main circle, as shown in Figure 9.8. The movement of the smaller circles expressed the broadening of Modex's original business model from skin to related indications. This graphic expansion thematically represented opening a pipeline to potential new revenue sources. This Bumper slide also established the agenda for the material to follow.

In his narration, Jacques described how he planned to cover each of the skin indications in the same order. If this sounds familiar, it is a composite of the Form/Function and Parallel Tracks Flow Structures, expressed graphically.

Then, using highlighting (see Figure 9.9), Jacques moved into the first indication, Chronic Wound.

Jacques then drilled down into the Chronic Wound topic for a total of 18 slides, most of them as simply constructed as the Bumper slide. For instance, he used a pyramid to describe the various layers of the market. The pyramid appeared in a five-slide sequence, and again later within the segment as a three-slide sequence. Two of the slides were pictures, two were bar charts, and two were boxes with minimal text, laid out to indicate related information. Only four of the 18 slides in this entire sequence were text. This was certainly not a Presentation-as-Document.

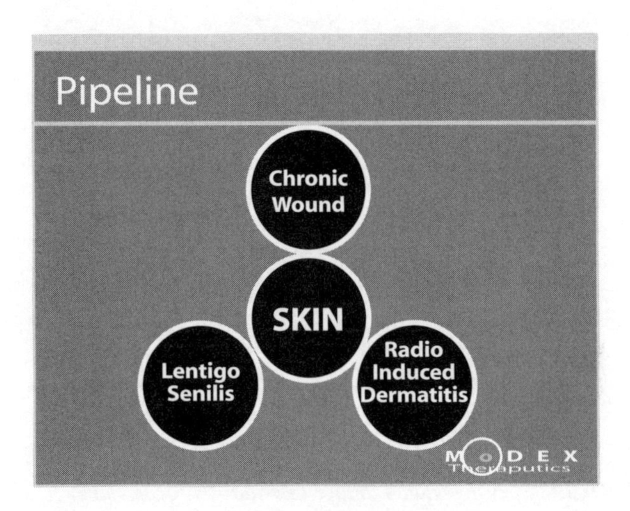

Figure 9.8 *Modex Therapeutics' expanded business focus slide.*

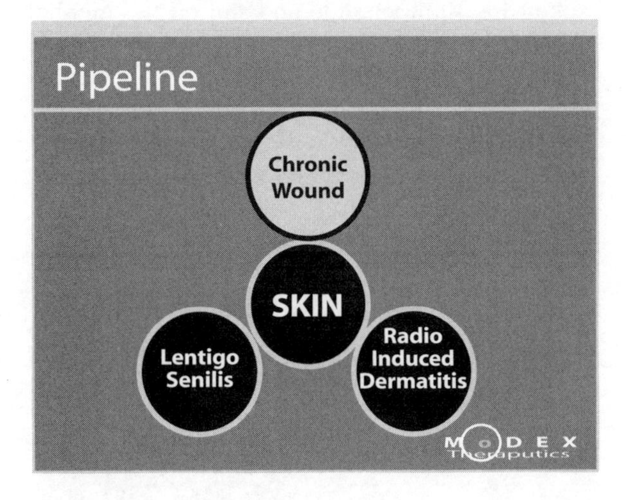

Figure 9.9 *Modex Therapeutics' Chronic Wound focus slide.*

Now Jacques was ready to shift to Radio Induced Dermatitis, and he did this by returning to his Bumper slide and shifting the highlighting to the new circle, as shown in Figure 9.10.

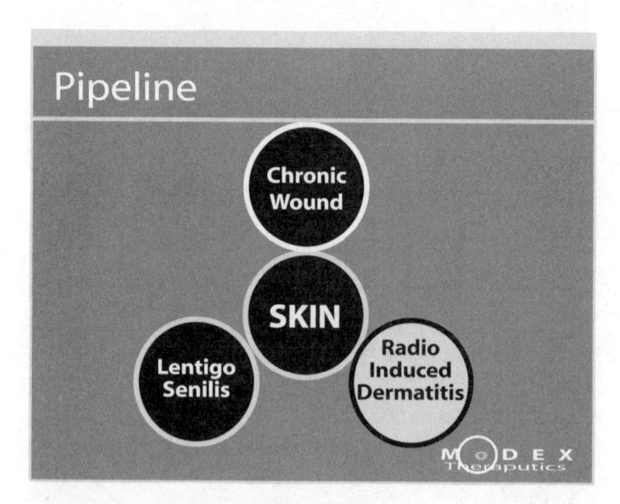

Figure 9.10 *Modex Therapeutics' Radio Induced Dermatitis focus slide.*

He covered the Radio Induced Dermatitis topic in a five-slide sequence. Then he came back to his Bumper slide and shifted the highlighting to the last indication, Lentigo Senilis, as shown in Figure 9.11.

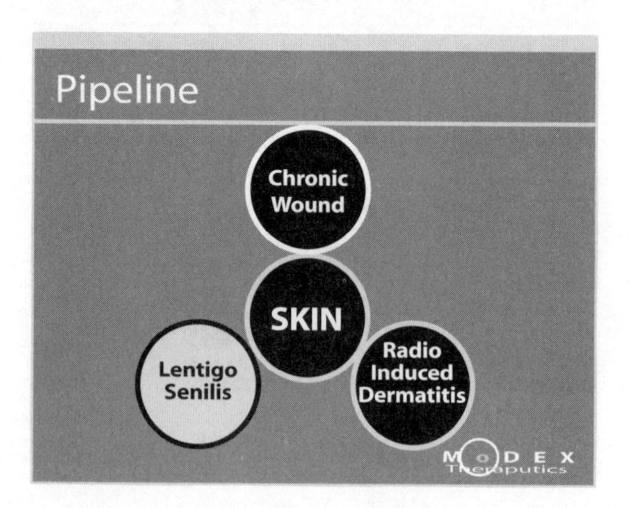

Figure 9.11 *Modex Therapeutics' Lentigo Senilis focus slide.*

Note that there are a sufficient number of slides between Bumper slides to warrant their repetitive use without appearing patronizing.

Jacques covered Lentigo Senilis in another five-slide sequence, two of which were pictures and three of which were text. By then he was done with his core business and was ready to move into the home stretch of the presentation.

While Modex's story is a complex one with many supporting details, Jacques' use of consistent graphics, linked and united by Bumper slides, made the story clear and easy to follow.

2. Indexing/Color Coding

Indexing/Color Coding is the second Graphic Continuity technique that helps you navigate through longer presentations. While Bumper slides appear intermittently, Indexing uses a recurring object that is highlighted with different colors or shades coded to map the different sections of your presentation. This approach helps your audience track your flow with a minimum of effort on their part . . . as well as yours.

This technique is often used in extensive technical presentations that have multiple sections, each of which is covered in deep, granular detail. The best way to portray this technique is with the simple example of a four-part strategy, depicted in Figures 9.12 through 9.15.

This company's four-part business strategy is represented by a pie icon divided into four different-colored wedges, as shown in Figure 9.12. If you were the presenter, you would show this slide near the start of your presentation to describe and discuss your overall strategy. When focusing on the first strategy element (Strategy A), you would shift the pie icon to the upper-right corner and highlight one wedge, using its signature color, as shown in Figure 9.13. You would then add details about this part of the strategy in the bullets. As a result, the pie becomes clearly identifiable as an index.

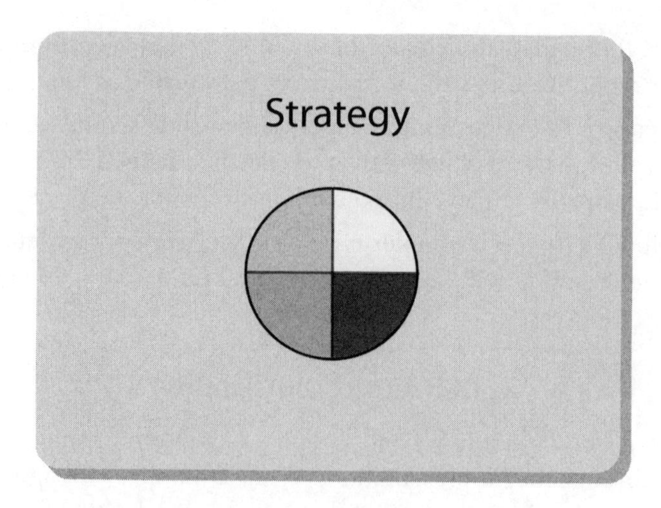

Figure 9.12 *Indexing/Color Coding a four-part business strategy.*

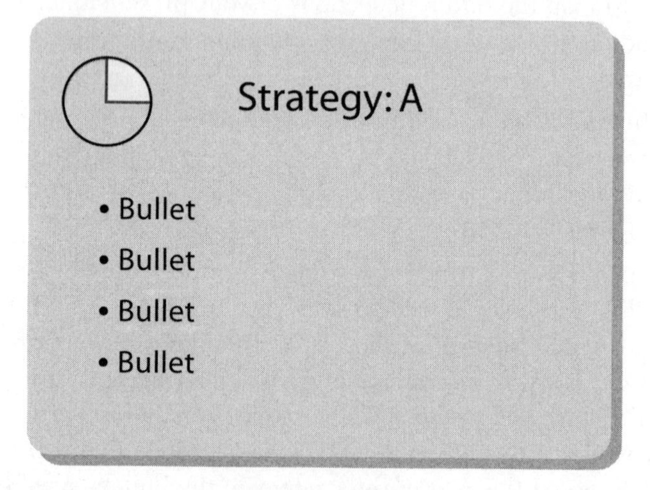

Figure 9.13 *Highlighting the first strategy element.*

The next slide (see Figure 9.14) moves on to the second part of the strategy. The highlight in the index shifts to the second pie wedge and its signature color. The explanatory bullets change, providing you with discussion points on which to elaborate.

You can already anticipate the progression of the next two slides. The highlight will shift to each of the next two wedges in turn, coming around the horn to complete the circuit. Using the Indexing/Color Coding scheme makes it easy for your audience to stay with your flow. It also creates the subtle, subliminal psychological dynamic of raising expectations and meeting them. Your audience *knows* that the third and fourth pie wedges will soon appear on the screen, and when the color shifts, exactly as they expected, they feel a sense of resolution and satisfaction.

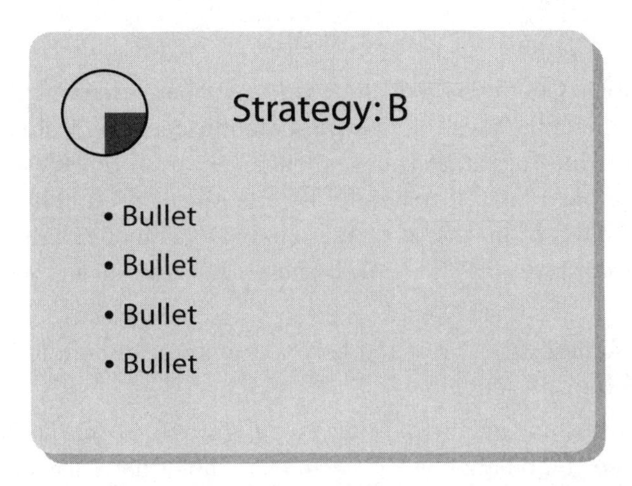

Figure 9.14 *Highlighting the second strategy element.*

Now let's jump ahead to the end of the circuit, where, in Figure 9.15, all four parts of the strategy are reprised in the large pie, with all the colors at their full original value. With this slide, you could summarize the strategy and then, by extending the pie with a ring, go on to describe future strategic directions, thus adding value.

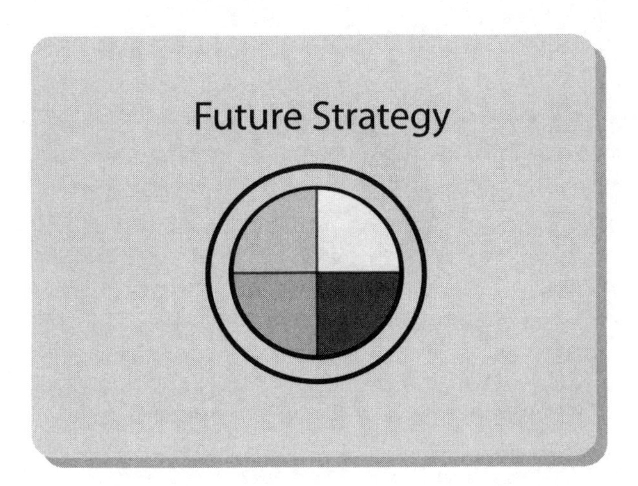

Figure 9.15 *Strategy recap.*

This sequence of slides serves as a series of guideposts, visually supporting your narrative as you *navigate* through all the stages of your presentation and lead your audience to your conclusion.

3. Icons

The third Graphic Continuity technique is *Icons*. These are symbols that express relationships among the ideas of your presentation. Icons indicate these connections in a shorthand form that your audience can grasp quickly and intuitively. Since Icons are pictorial, they meet the Less Is More criterion by default. Icons also provide relief from the Presentation-as-Document Syndrome, with its extended series of text-only slides. Well-chosen Icons make an excellent presentation tool.

Here are some classic Icons and how you might use them to help illustrate the content of your presentation:

In Chinese cosmology, yin and yang (see Figure 9.16) are the dual, contrasting forces whose combination gives rise to everything that exists in the universe. For Western audiences, the yin/yang Icon simply means "two-part harmony." You might use a version of this Icon in a presentation whose theme is the integration of two forces: for example, when explaining the benefits of a merger between two companies or when presenting your firm's two product lines and the complementary benefits they offer.

Figure 9.16 *The yin/yang Icon.*

Many people subscribe to what is commonly known as the "rule of threes," in the theory that three is an easy number to remember. If your story contains three key elements, you can pick one of a variety of triplex Icons to express them: the triangle, the trident, the three-legged stool, three interlocking arrows, or the three-ring Icon, as shown in Figure 9.17.

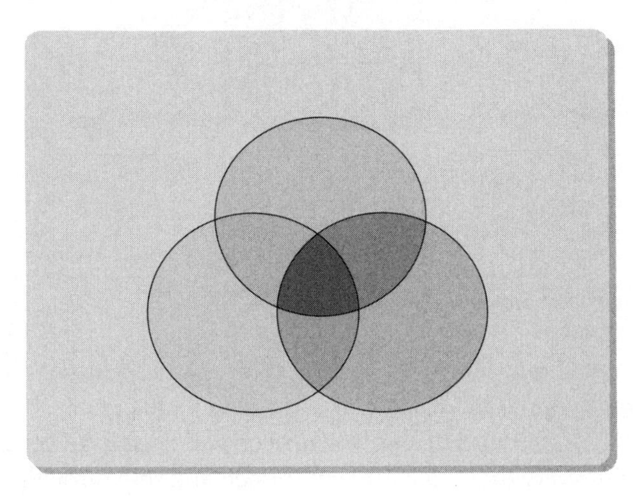

Figure 9.17 *The three-ring Icon.*

This image represents harmony among three items. You might use a version of this Icon when describing how three forces, groups, or units work together or combine in some way. For example, you could use it when presenting the story of a company that serves three distinct but overlapping markets or customer groups, such as businesses, government, and schools, or when describing a company's three-pronged competitive strategy. If you chose this Icon, be sure to set two rings at the bottom, creating a stable image.

If your story contains four key elements, try the Icon shown in Figure 9.18, which suggests four-part harmony. You might use this in a presentation to describe four complementary products, four departments that work together (Sales, Marketing, Information Technology, and Customer Service), four elements in a strategic plan, or four markets in which a company operates (North America, South America, Asia, and Europe).

Figure 9.18 *A four-part harmony Icon.*

The pyramidal Icon shown in Figure 9.19 suggests a set of hierarchical relationships. It could represent layers of management in an organization; an array of products, ranging from those that are low-priced and mass-marketed (the bottom layer of the pyramid) to those that are high-priced and sold through specialty outlets (the top of the pyramid); or a sequence of concepts that build on a broad foundation and then on one another, culminating in the most advanced concept.

Figure 9.19 *A hierarchical relationship Icon.*

In Chapter 4, "Finding Your Flow," you saw how a pyramid expressed a Spatial Flow Structure (please refer to Figure 4.1). In my own presentations and workshop programs, I describe the components of an effective presentation by starting with a solid story at the foundation of the pyramid, and then build up through each tier: graphics, delivery skills, presentation tools, and on to the top with Q&A skills. Think of the dozens of other Icons you've encountered, each with its own visual statement that can help illustrate your story.

Some companies, like Cisco, devote considerable effort to creating Icons. Given its complex technology, Cisco endeavors to make comprehension easier for its audiences by expressing that technology in images rather than text. *A picture is worth a thousand words.*

When Cisco launches a new product family, it carefully creates a specific graphic image to represent that family. The company expends even more effort to develop conceptual graphic images that represent industry-wide functions and processes. These conceptual Icons are so important that Cisco has established a set of specific guidelines for their creation.

Gary Stewart has been a technical illustrator for Cisco since 1991. He has overseen the creation of Icons that represent such universal networking technology elements as the router, the ATM switch, the network cloud, the firewall, the bridge, and the communication server, all shown in Figure 9.20.

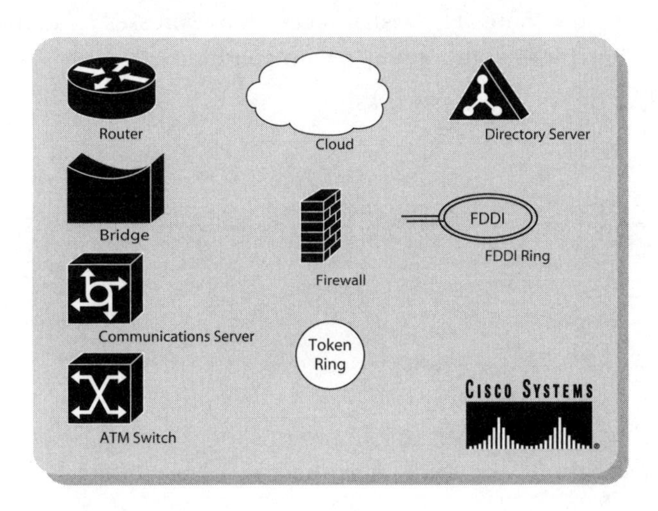

Figure 9.20 *Cisco Systems' lexicon of Icons.*

Over the years, Cisco has accumulated a full lexicon of Icons that they use in their presentations and literature. Cisco also makes these Icons available on a public domain site (www.cisco.com and search for "icons"). As a result, the symbols have become industry nomenclature as well as an enormous branding initiative for Cisco.

If there's an Icon that captures and expresses the relationships among your ideas, consider using it throughout your presentation as a visual symbol for your audience.

4. Anchor Objects

The fourth Graphic Continuity technique is called *Anchor Objects*, in which a recurring image appears in a succession of slides to express a continuing relationship. Where Index/Color Coding creates continuity with a recurring icon, such as a pie or a pyramid, the Anchor Object option creates continuity with a recurring image that is an integral part of the illustration. This image or object can be a photograph, a sketch, a map, an icon, a screen shot, a logo, or clip art.

Here, you begin a series of slides in which the first in the series contains the Anchor Object. In the next slide, this object remains in place and is joined by another object, establishing a relationship between the two objects. In the next slide in the series, the two objects remain anchored in place while still another object appears to join them. This visual succession expresses relationships among the objects and provides your audience with continuity.

Figure 9.21 shows how this works.

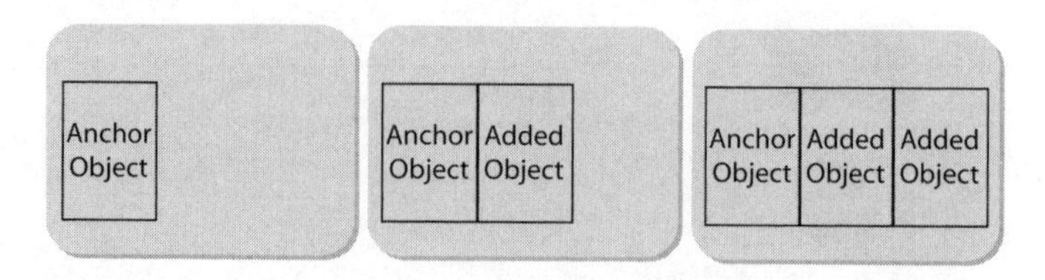

Figure 9.21 *Anchor Object progression.*

As you bring in new objects to join the Anchor, you can also create emphasis by highlighting with different colors or shading, or by adding borders. Another way to create emphasis is by enlarging any one object as a callout, or by adding depth with foreground or background elements. This next example illustrates this.

In Chapter 4, you read about Dr. Robert Colwell, one of the leaders of the engineering team that developed Intel's next-generation integrated circuit, the P6. For the launch presentation at a technical symposium, Bob used a schematic diagram of the new chip, in basic black and white, as the Anchor Object to track the flow of his presentation.

In the slides shown in Figures 9.22 through 9.26, each slide is a visual variation on the one before. Building on the schematic diagram, Bob highlighted some of the objects by reversing the black and white, thus shifting the audience's focus from one element of the chip's design to another. As Bob navigated through his presentation, he also used enlarged callouts in the foreground.

This simple graphical design provided a clear road map for a highly technical narrative. By the end of Bob Colwell's presentation, not only did the audience of his peers have a clear understanding of all the key features of the new Intel chip design, they had gained a healthy respect for Bob and also for Intel.

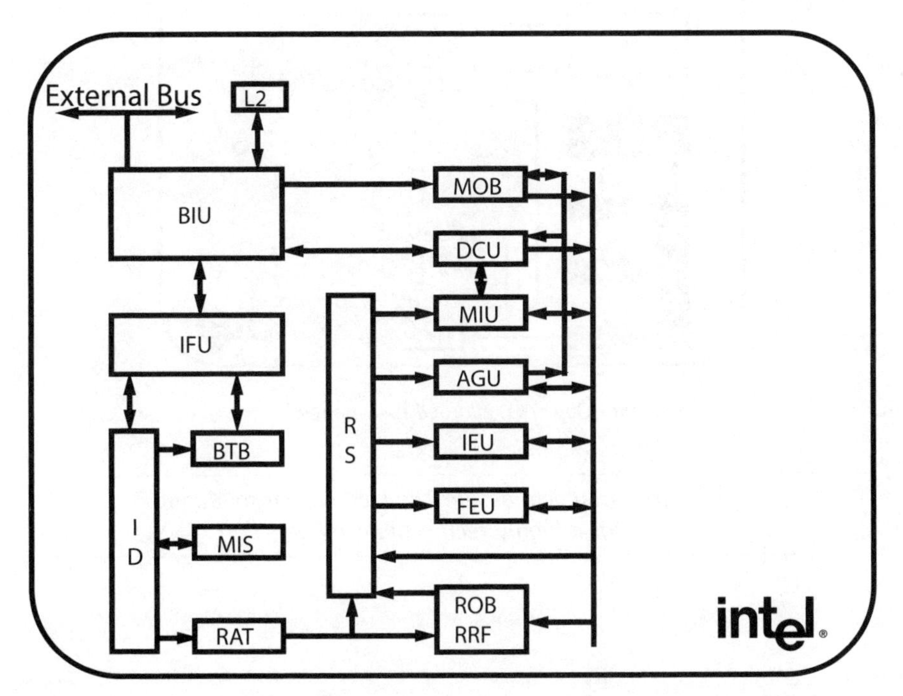

Figure 9.22 *Intel uses the bus diagram of a new chip design as an Anchor Object.*

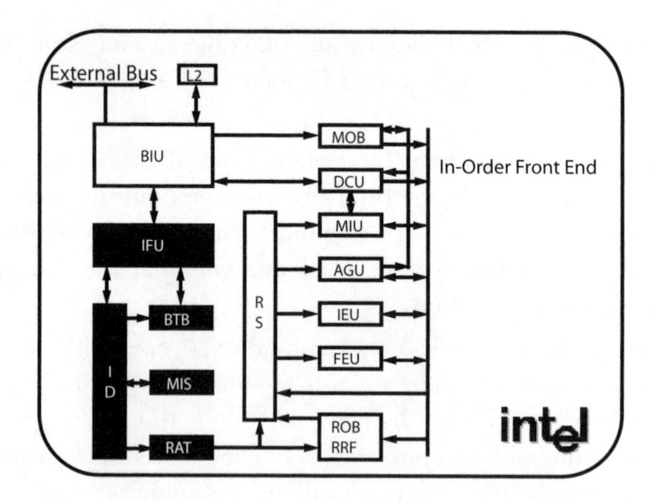

Figure 9.23 *Part of the Anchor Object is highlighted for emphasis.*

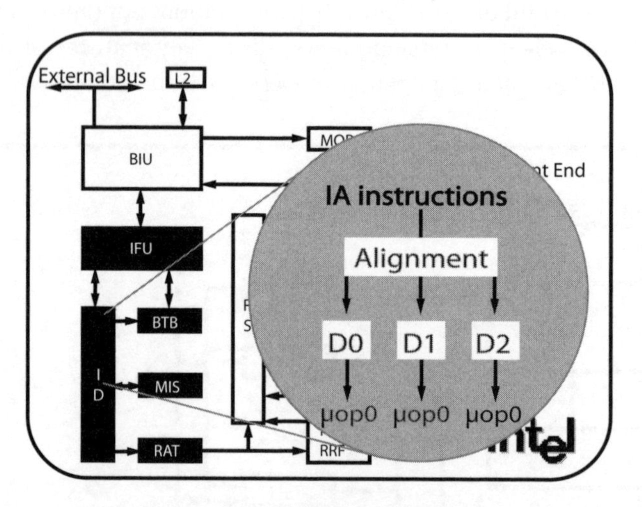

Figure 9.24 *Part of the Anchor Object is enlarged for emphasis.*

This simple graphical design provided a clear road map for a highly technical narrative.

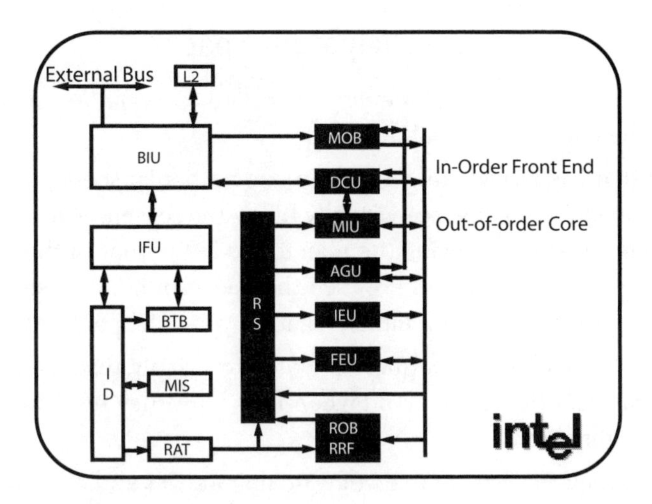

Figure 9.25 *The highlighting is shifted.*

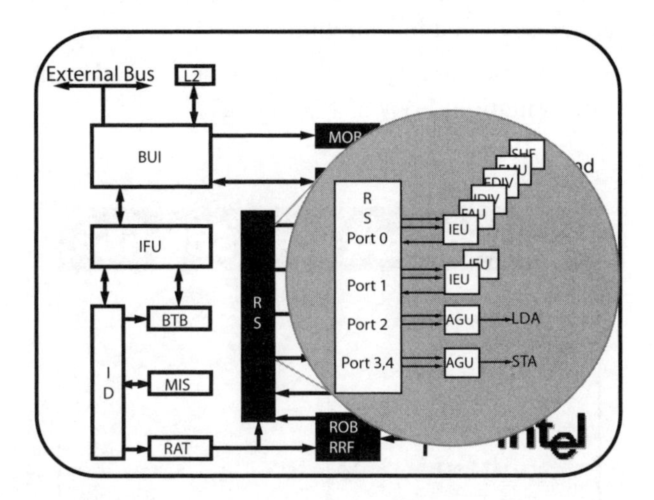

Figure 9.26 *Another part of the Anchor Object is enlarged for emphasis.*

Remember Bob's postmortem: "I didn't get any of the hardball questions I was worried about . . . I wonder if maybe potential hostile questioners were holding back, for fear of 'losing' the argument to somebody who appeared to have done his homework and seemed thoroughly in control of the proceedings." Bob's graphics certainly contributed to that perception.

5. Anticipation Space

The final Graphic Continuity technique is called *Anticipation Space*, and it is drawn from the cinema. Here's how it works:

Imagine that you're watching a film. In one particular scene, one of the characters is seen in close-up, his face virtually filling the screen. After a moment, the camera widens out slowly, leaving the man in the left corner of the screen. While the man remains visible, we can now see the background. The setting is a cafe, and the man is sitting alone at a table. What do you anticipate?

When I ask my clients this question, they invariably respond, "Someone or something will enter the scene." And when I ask why, they say, "To fill the empty space on the right side."

Of course, they're correct. Years of watching movies and television have conditioned us to recognize that when an empty space appears on the screen, that space will be filled. In our imagined movie scene, the man's lover would appear and take the seat across the table from him.

The same sense of audience anticipation can be created using simple graphics in a business presentation. Look at Figure 9.27.

Requirement	Our Solution
Item	
Item	
Item	
Item	

Figure 9.27 *Anticipation Space.*

What do you anticipate when you look at this slide? Of course, you expect the empty space on the right side of the slide to be filled. More specifically, you expect *four* items to appear, to correspond to the four items on the left side. Four is the only possible number. Three items would disappoint you; five would confuse you.

The *only* possible slide to follow Figure 9.27 is the one shown in Figure 9.28. This slide answers our questions, satisfies our expectations, and fulfills our anticipation. The subliminal message: *Our company promises, and it delivers.*

Requirement	Our Solution
Item	Item
Item	Item
Item	Item
Item	Item

Figure 9.28 *Anticipation fulfilled.*

Anticipation Space is a simple but visually powerful technique, and another instance of subliminally managing your audience's minds.

Here's how one company used Anticipation Space to illustrate an Issues/Actions/Matrix/Parallel Tracks Flow Structure: The company had faltered and was attempting a turnaround. It brought in a new management team that decided to reposition the entire business model. They called on me to help them develop a presentation to describe the turnaround strategy to their employees, investors, and other concerned stakeholders.

The new team identified five factors they had to address to reposition the company. For their first slide (see Figure 9.29), they designed a matrix that listed the five factors on one side, leaving corresponding blank spaces on the other side for the actions they planned to take, setting up the anticipation.

In the next slide (see Figure 9.30), they filled the empty spaces with an action item for each factor. Thus, the anticipation was resolved.

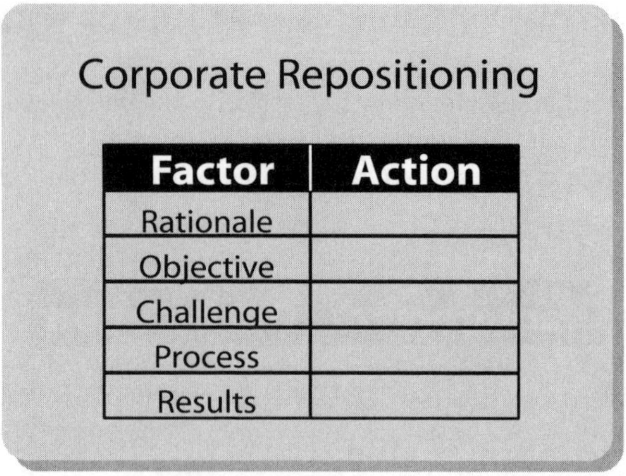

Corporate Repositioning

Factor	Action
Rationale	
Objective	
Challenge	
Process	
Results	

Figure 9.29 *Anticipation Space with the Matrix/Parallel Tracks Flow Structure.*

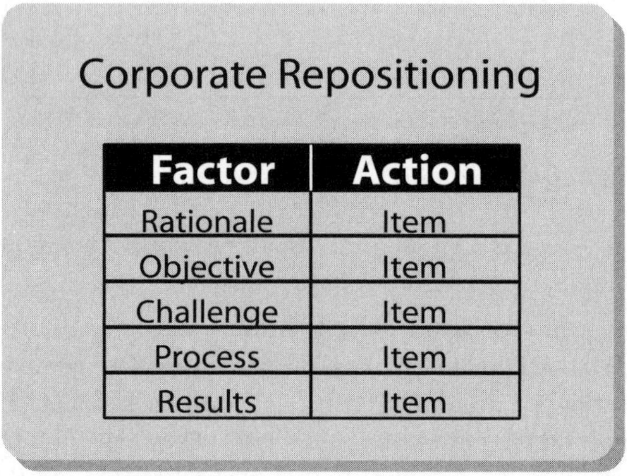

Corporate Repositioning

Factor	Action
Rationale	Item
Objective	Item
Challenge	Item
Process	Item
Results	Item

Figure 9.30 *The same matrix, with the Anticipation Space filled in.*

Using the same format, they continued on to another slide to discuss their corporate identity. They listed the factors they had to address on the left and then filled in the spaces on the right with actions they intended to take. On a third slide, following the same Parallel Track, they used the same format again to discuss their product positioning. Again, they listed the factors on the left and filled the spaces on the right with the actions they intended to take.

Those three slides, flanked by an Opening Title slide and a Closing Title slide, served as the basis for their *entire* presentation, as shown in Figure 9.31.

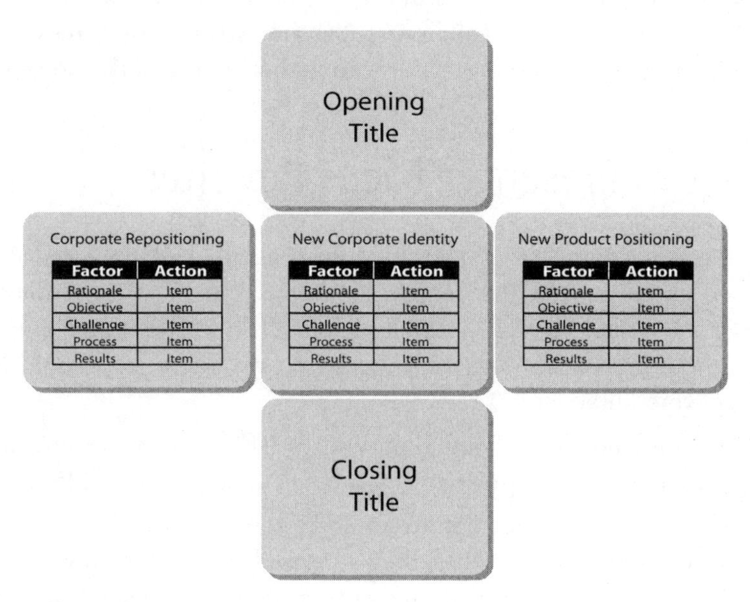

Figure 9.31 *The entire presentation.*

Simple? Yes, especially when you consider that the new team had arrived for their session with me with *48 all-text slides*! The revised five-slide presentation served as a simple and vivid snapshot of their plan to execute their turnaround strategy.

To recap, the five Graphic Continuity techniques are:

1. **Bumper slides** are the graphic dividers inserted between major sections of this presentation to serve as clean, quick, and simple transitions.
2. **Indexing/Color Coding** uses a recurring object as an index, highlighted in different colors to map the different sections of a longer presentation.
3. **Icons** express relationships among ideas with recognizable symbolic representations.
4. **Anchor Objects** create continuity with a recurring image that is an integral part of the illustration.

5. **Anticipation Space** uses empty areas that are subsequently filled, setting up and then fulfilling subliminal expectations.

These five simple Graphic Continuity techniques can have a major impact on the clarity and coherence of your presentations. Artfully used, they can make even a long, complicated story easy for your audience to absorb and remember.

Presenter Focus Revisited

Let me pose a question. Throughout this entire chapter, and in the previous three chapters on graphics, I haven't mentioned Point B or the WIIFY once; while, throughout the earlier part of this book, I repeatedly stressed their importance. Where in the Storyboard, or on which slide, do you think you should put your Point B and your audience's WIIFY?

The answer is *nowhere*. Point B and the WIIFY don't appear on *any* slide. They are stated by you, the presenter. *You* tell your story; your slides do not.

I asked this as a trick question to reinforce the concept of Presenter Focus. The slides are not the presentation; they are simply presenter support. It is you, the presenter, who must *grab* your audience at Point A, *navigate* them through all the parts, and *deposit* them at Point B.

Imagine a company CEO, in a presentation to an audience of potential investors, showing the slide in Figure 9.32.

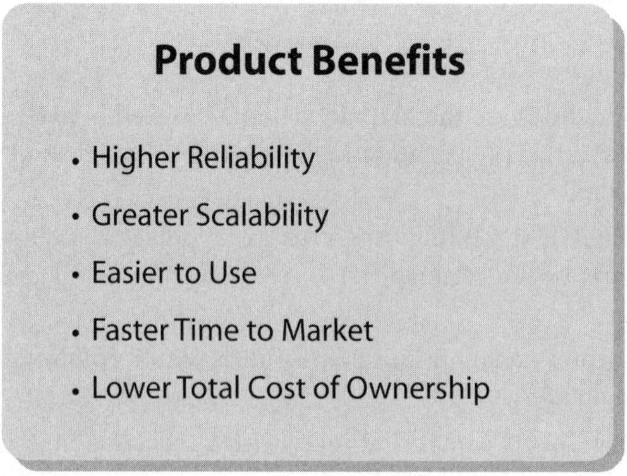

Product Benefits

- Higher Reliability
- Greater Scalability
- Easier to Use
- Faster Time to Market
- Lower Total Cost of Ownership

Figure 9.32 *Where's the WIIFY?*

After discussing the various product benefits listed in the slide, the CEO could summarize by saying, "You can see that our product provides a rich set of benefits to our customers," and then move on to the next slide. That, however, would be an opportunity missed.

Instead, the CEO should add: "This rich set of customer benefits produces repeat business for our company. Repeat business translates into recurring revenues. Recurring revenues translate into shareholder value, and an excellent investment opportunity."

That statement contains both the WIIFY and Point B, *neither of which appears on the slide*. By coming from the presenter, the presenter leads the audience to a conclusion. That conclusion is *Aha!*

Graphics and the 35,000-Foot View

Let's recap what you've learned about presentation graphics in the past four chapters:

The overarching principles that govern *all* graphics are Presenter Focus, Less Is More, and Minimize Eye Sweeps. All of these principles are particularly important in creating text and numeric slides.

Mix and match *text* and *numeric* slides with the other two major graphic options, *pictorial* and *relational* slides, to avoid the dreaded Presentation-as-Document Syndrome. Use all these elements to craft the slides that will express your story in a Storyboard Form.

The Storyboard is the 35,000-foot panoramic view of the entire presentation. Microsoft PowerPoint's Slide Sorter view provides this panoramic aspect. The Storyboard enables you to check that the graphics convey the sequence clearly. If the transitions from slide to slide or section to section aren't clear, consider reorganizing your slides, or crafting strong verbal transitions to make the logic apparent in your narration.

Finally, use the Graphic Continuity techniques, *Icons*, *Indexing/Color Coding*, *Anchor Objects*, and *Anticipation Space*, to help navigate your audience through your presentation. Use *Bumper slides* to help the flow by creating clear transitions between topics.

These tools can ensure that everyone in your audience understands exactly where he or she is at all times, as well as how any given slide fits into the overall flow of the presentation.

Figure 9.33 shows the Storyboard Flow Form. It combines the Story Form (from Chapter 5, Figure 5.2) with a Storyboard, relating all the preceding techniques in one integrated view. By combining a well-constructed story that has a clear beginning, middle, and end with the slide layout, you can see how to express your entire presentation. This is the ultimate *forest view*.

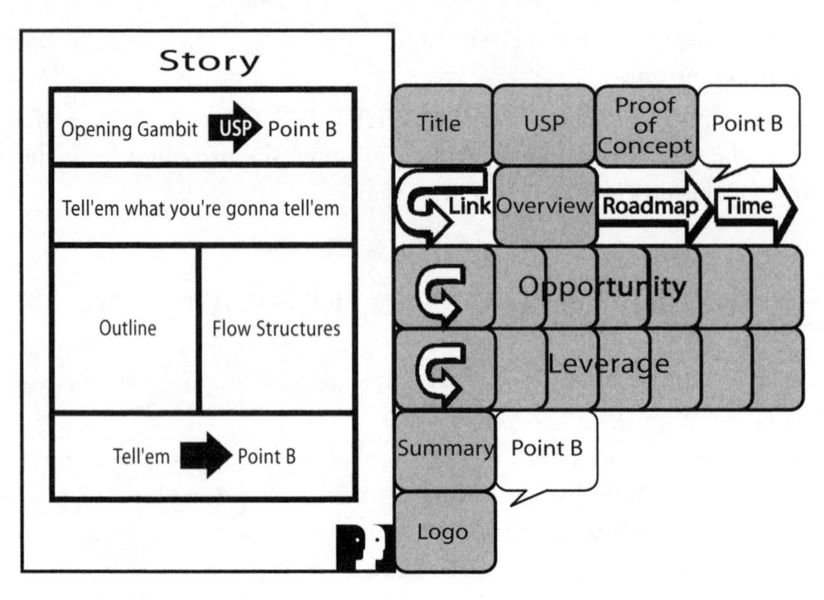

Figure 9.33 *The Storyboard Flow Form.*

By combining a well-constructed story that has a clear beginning, middle, and end with the slide layout, you can see how to express your entire presentation.

The form provides a useful guide for putting together all your essential elements. Notice that I've left the slides blank. They contain only high-level concepts and dynamic inflection points so that you can focus on the flow. Here's how it works:

- **The Opening Title slide** is the background for the Opening Gambit, which links to:
- **The Unique Selling Proposition (USP) slide**, which links to:
- **The Proof of Concept slide**, where the linkages culminate in a spoken Point B.

- This is followed by the all-important (and often overlooked) forward link to the **Overview slide**, which serves as a preview of the entire presentation. This is followed by another forward link to a statement that the Overview serves as a *road map*, or an agenda, which is then followed by a forecast of the time. Then, we return to the top of the road map to begin the presentation.

The narration then navigates through all the slides in the full body of the presentation. In the example in the figure, the Opportunity/Leverage Flow Structure organizes all the slides. Upon concluding the body, move to:

- **The Summary slide**, which recapitulates the entire presentation and concludes in a spoken Point B.
- **The Company Logo slide** leaves the audience with the brand image of your company, reinforcing your Point B (*I see and I remember*).

Here's how Carol Case, the salesperson from Argus Insurance, applied the Storyboard Flow Form to combine her narrative with her slides:

- **Over her Title slide**, Carol delivered her Opening Gambit anecdote about the Argus customer whose home burned down. Then she continued with her link: "Our customer is like many customers who purchase a basic policy not customized to their individual needs. He now realizes that means being just one step away from disaster. Luckily, we at Argus Insurance have a solution."
- **Clicking to her USP slide**, Carol continued: "Argus can provide you with a customized, value-added package of insurance that provides for your individual needs to protect you against serious financial loss." She linked to her Proof of Concept over the same slide: "Maybe that's why Argus is one of the fastest-growing insurance brokers in the state." Continuing with the same slide, she delivered her spoken Point B: "I know that you'll want to take advantage of the opportunity and sign up for this important coverage today."
- **Clicking to her Overview slide**, Carol then linked forward from Point B: "So that you can consider signing with Argus . . . " Then she ran through the bullets of her Overview slide and concluded by saying: "Please consider this a road map for the next 15 minutes of my presentation. Let's begin at the top." Please notice that Carol did not refer to her Overview as a road map until the end of the slide, thus making the Overview an extension of her Point B as well as adding further continuity.

The Storyboard Flow Form provides the 35,000-foot overview of your presentation, which in turn enables you to convey a clear sense of continuity and flow to your audience. When you feel in command of your material, you communicate a sense of confidence to your audience and heighten the power of your presentation.

Chapter Ten

Bringing Your Story to Life

Company Example

- **Central Point Software**

The past few chapters have focused on the visual side of your presentation. Now it's time to take a big step back and think again about your presentation as a whole, particularly about its verbal side.

Let's return to the Story Form, which you first saw in Chapter 5 (refer to Figure 5.2). This single form contains the basic building blocks of any presentation story: the Opening Gambit, Point B, the Outline, and the Flow Structures.

The form lays out the overall flow of your presentation story, but because it is at a high level of abstraction, the forest view, it shows only the most salient points. Your presentation comes to life only when you add granularity: the specific verbiage you will use to tell your story.

How do you go from the bare bones presented in the Story Form to a complete presentation? The answer is through a process of preparation and practice called *Verbalization*.

Verbalization: The Special Technique

Verbalization means turning your outline into a full-fledged presentation by practicing it beforehand. Speak the actual words you will use in your presentation *aloud*, accompanied by your slides. Do it just the way you will do it when you are in front of your intended audience. A truly effective presentation is practically impossible without this special technique.

Yet many businesspeople are reluctant to Verbalize. Some claim that they can't verbalize until the presentation is "baked." Others feel self-conscious or uncomfortable about "performing" in front of others. Still others view Verbalization as too elementary. Whatever the reason, they try to short-circuit the process, often saying, "Oh, don't worry. I'll rehearse my presentation before I have to deliver it, and it'll be just fine."

Unfortunately, this is how most businesspeople rehearse their presentations: As the slides flip by on the screen, the presenter glances at each one and says something like, "Okay, with this slide I'll say something about our sales revenues . . . and then with this slide I'll say something about our path to profitability . . . and then with this next slide I'll show a picture of our lab and talk a little about R&D."

Sound familiar? As a form of rehearsal, it is completely unproductive. Talking about your presentation is not an effective method of practicing your presentation, any more than talking about tennis would be a good method of improving your backhand. This is known as *disembodiment* because it distances the presenter from the presentation.

A close cousin of disembodiment, and an even more common practice, is mumbling. We've all seen it done. The presenter either clicks through the slides on the computer or flips through the pages of a hard copy of the slides while muttering unintelligible words.

Neither of these methods is Verbalization.

The *only* way to prepare a Power Presentation is to speak it aloud, just as you will on the day of your actual presentation. By talking your way through your entire presentation in advance, by articulating your key points and making the logical connections, by doing a real-time Verbalization, you'll crystallize the ideas in your mind.

The only *way to prepare a Power Presentation is to speak it aloud, just as you will on the day of your actual presentation.*

Here's a personal perspective: I present nearly every business day of my life, so I don't have to Verbalize material that I've delivered in one form or another thousands of times before. When I have to present new material, however, I Verbalize it extensively in advance. An example is the pyramid presentation in Chapter 4 (refer to Figure 4.1). You'll recall that I often give this presentation at conferences, where I describe the essential components of any presentation by starting with the story as the foundation at the base of the pyramid.

Even though I've given variations of this presentation countless times, I still Verbalize each new iteration a *dozen times or more* in advance. When I have to deliver completely new material, I *double* that number of repetitions.

Another perspective is the example of the CEO of a start-up technology company. The man, who began his professional career as a scientist, had developed his esoteric technology in his garage and then bootstrapped his company without ever having to make presentations. But when his technology took off and his company was about to go public, he knew that he would have to stand and deliver the IPO road show. He retained my services, and we went through all the steps you've learned in this book, except for Verbalization.

On the morning of the day he was to deliver his road show to the investment banking team that would be selling his offering, he panicked. I printed his slides on paper, spread them out on a conference room table, and asked him to talk through them. But he stammered as he did, and his panic intensified. I asked him to start again. As he worked his way through the slides the second time, he stammered less and less. I asked him to do it again and again.

By the fifth run-through, his stammering had disappeared; by the sixth, he was beginning to develop continuity; by lunchtime, when the bankers arrived, he was able to deliver a positively fluid presentation. The solid foundation of the well-developed story and graphics combined with Verbalization gave the apprehensive presenter the comfort level he needed.

The solid foundation of the well-developed story and graphics combined with Verbalization gave the apprehensive presenter the comfort level he needed.

In the Power Presentations programs, I step my clients through the process of developing a compelling start to their presentations with the brisk 90-second sequence that you read about in Chapters 5 and 9: the Opening Gambit, the USP, a forward link to Point B, another forward link to Tell 'em, and one more forward link into the Overview. Invariably, the first time they try this new sequence, they struggle and stumble. I ask them to do it again. The second time, they struggle less, and still less each succeeding time they Verbalize. Try it yourself, and you'll see how readily this powerful method works.

Each time you Verbalize your presentation, and each time you deliver it before a live audience, you should expect the words of your narrative to vary slightly. The logic of your well-honed outline, and your improved comfort level, will guarantee that every time you deliver your presentation, you'll express your key ideas clearly and persuasively. That's the power of Verbalization.

Spaced Learning

Educators distinguish between distributed learning and massed learning. Distributed learning, occurring over time, is the more efficient method, because it allows for absorption and understanding. Distributed learning is a synonym for *Spaced Learning*; massed learning is a synonym for cramming.

Every schoolboy and schoolgirl in America has heard the story about how Abraham Lincoln wrote the Gettysburg Address on the back of an envelope. The implication is that he dashed off the classic speech with little or no advance preparation.

This tale has several variations. In his Pulitzer Prize-winning book, *Lincoln at Gettysburg*, Garry Wills describes other lesser-known versions of the same story that have Lincoln "considering [the speech] on the way to a photographer's shop in Washington, writing it on a piece of cardboard . . . as the train took him on the 80-mile trip . . . penciling it . . . on the night before the dedication, writing it . . . on the morning of the day he had to deliver it, or even composing it in his head as Everett [the prior speaker at the Gettysburg ceremony] spoke."[1]

Maybe the appeal of these stories comes from the fact that so many businesspeople prepare their *own* presentations in this fashion. Charles M. Boesenberg, now a member of the Board of Directors of Callidus Software and Interwoven, was the president of MIPS, a semiconductor design company, and a participant in the road show for its public offering in 1989. Later, when Chuck became the CEO of Central Point Software, he retained my services for that company's IPO road show. I introduced Chuck to the same story techniques you've been learning here: the Framework Form, Brainstorming, Clustering, and Flow Structures.

Halfway through the first day of the program, Chuck broke into a big grin and said, "At MIPS, we did *all* of this in the taxi on the way to the airport to begin our road show!"

Chuck is not alone. Whenever I ask my clients if they ever whip together their own presentations at the last minute, I usually get a round of sheepish grins. Their variations on the theme include:

- "Oh, I didn't have time to do my presentation, but I'll wing it!"
- "I'm running late in a meeting. Can you deliver my pitch for me?"
- "I won't have time to work on this, so have Marketing put something together."

[1] Wills, Garry. *Lincoln at Gettysburg: The Words That Remade America* (New York: Simon & Schuster, 1992).

Finally, there's the approach that, unfortunately, far too many businesspeople use: "Let's see . . . I can use three of Tom's slides, six of Dick's, and four of Harry's." Remember that the name for this method is "Frankenstein body parts."

If any of those approaches sounds familiar, it might make you feel better to believe that Abraham Lincoln produced his masterful Gettysburg Address in much the same way. However, just the opposite is true. Garry Wills puts to rest what he calls that "silly but persistent myth" by detailing the complex history of Lincoln's creation of the Gettysburg Address. Wills describes its antecedents in classical rhetoric, a subject that was popular in the 19th-century; Lincoln's own immersion in literature and the Bible; a lifetime of study and practice; and his admiration for the oratory of Daniel Webster. Wills adds about Lincoln, "He was a slow writer, who liked to sort out his points and tighten his logic and phrasing. This is the process vouched for in every other case of Lincoln's memorable public statements."

Just for good measure, Wills also tells us that Lincoln combined Verbalization with Spaced Learning: "This surely is the secret of Lincoln's eloquence: he not only read aloud, to think his way into sounds, but wrote as a way of ordering his thoughts."

Wills then puts the improvisation myth about the Gettysburg speech completely to rest. He cites several specific pieces of historic evidence indicating that Lincoln organized his information and ideas in Washington at least two days before the speech and continued to work on the text at multiple points along the way to the Gettysburg site. Then, on that memorable day of November 19, 1863, Lincoln rose "with a sheet or two" to deliver it. All of this for a speech that was just 272 words long!

From the sublime oratory of Abraham Lincoln to the mundane hobby of puzzles: Veteran crossword fans know that when they become stuck with a puzzle, they put it aside briefly. When they return to it a while later, they are unstuck. Presenters who recognize the value of Spaced Learning report that when they revisit their presentation or speech after a brief interval away from it, they can quickly see half a dozen ways to improve it. This is the result of seeing the presentation from a new viewpoint, or perhaps the result of letting the unconscious mind process the material while the conscious mind is otherwise engaged. Moreover, the result of this fresh perspective creates a greater command of the presentation and makes the presenter feel . . . as well as appear . . . more confident.

With all of these advantages to Spaced Learning, why do so many presenters relegate preparing their presentations to the last minute? Most businesspeople, like you, are constantly overloaded, pressured, and rushed. However, if you accept the view that every presentation is a mission-critical event, then those excuses are not valid.

We're all familiar with Andy Warhol's comment that everyone will be world-famous for 15 minutes. I'd extend Warhol's observation to say that every presenter has 15 or 30 or 60 minutes . . . whatever time is allotted or assigned . . . to get the audience to *Aha!* Wouldn't you want *every single* one of those minutes to be all that it can be? Wouldn't you want to employ *every single* tool and technique at your disposal to *grab* your audience's minds at the beginning, *navigate* them through all the parts, never letting go, until you *deposit* them at Point B?

You can do it all through the power of Spaced Learning.

I practice what I preach. Most of the material in this book has evolved over the course of the 20 years since I started Power Presentations, except for the section you've just read. I decided to write it *after* I read the Wills book. It took me 37 drafts written over three days, not counting the additional drafts I did with input from my editorial consultant and publisher.

If you've benefited from reading this chapter, you can thank Spaced Learning. You too can use the same tool to your own advantage.

Internal Linkages

You'll be building your presentations using the key elements contained in the Story Form. Consider these elements as building blocks, and that these blocks need mortar to hold them together. The glue is a set of narrative tools called linkages: meaningful verbal transitions from one slide to the next and from one part of the presentation to the next.

The importance of linkages is best illustrated by returning to the investment banking conferences that I first described in Chapter 4. I attend these conferences because they let me observe many presentations in one place, in a short time. At each conference, I get to see dozens of executives, each of them experienced, poised, and confident speakers, make their management presentations. Yet the lion's share of them proceeds through each presentation in the same manner: they click to a slide on the screen and say, "Now I'd like to talk about . . . " and then they talk about that slide. Then the presenter clicks to the next slide and says, "Now I'd like to talk about . . . " and talks about that slide. A click to the next slide, and the presenter says, "Now I'd like to talk about . . . " and so on, throughout the presentation.

This trite phrase is not only a useless cliche, it also destroys any continuity or flow. In effect, it makes each slide start the presentation anew. This is the equivalent of *rebooting*, and it presents a major problem for the audience. With no context for the diverse ideas and no connection among them, the audience is stranded at the level of the trees, seeing only one tree at a time, with no idea of how one tree relates to another. The missing connections are the verbal glue that the presenter must provide.

These connections take two main forms: *Internal Linkages*, which are statements that tie together the various parts of your presentation, and *External Linkages*, which are statements that tie the various parts of your presentation to your audience. We'll cover the Internal Linkages, of which there are 12, in this chapter and save the External Linkages for the next chapter:

1. **Reference the Flow Structure.** Make repeated references to your primary Flow Structure as you track through your presentation.
2. **Logical Transition.** Close your outbound subject; lead in to your inbound subject.
3. **Cross-Reference.** Make forward and backward references to other subjects in your presentation.
4. **Rhetorical Question.** Pose a relevant question, and then provide the answer.
5. **Recurring Theme.** Establish an example or data point *early* in your presentation, and then make several references to it throughout your presentation.
6. **Bookends.** Establish an example or data point early in your presentation, and *never* mention it again until the end.
7. **Mantra.** Use a catchphrase or slogan repeatedly.
8. **Internal Summary.** Pause at major transitions and recapitulate.
9. **Enumeration.** Present related concepts as a group, and count down through each of them.
10. **Do the Math.** Put numeric information in perspective.
11. **Reinforce Point B.** Restate your call to action at several points throughout your presentation.
12. **Say Your Company Name.** State your company, product, or service name often.

Use these Internal Linkages liberally throughout your presentation, and make it easy for your audience to follow your flow. Let's look more closely at each of the 12 Internal Linkages, with brief examples of each.

1. Reference the Flow Structure

While all the other Internal Linkages are options that you can mix and match and use as needed, the first, Reference the Flow Structure, is not merely an option, but a necessity. It provides yet another way for you to bring your audience up from the level of the trees and give them a *continuing* view of the forest.

Here's how it works: If you've chosen Problem/Solution as the Flow Structure for your presentation, keep referring back to the problem your business is addressing. Throughout your presentation, use sentences such as, "So you can see how our unique solution addresses the problem that affects millions of people."

Similarly, if you've chosen the Opportunity/Leverage Flow Structure, keep referring back to the large opportunity and how your company's products or services are deploying to leverage it.

Or, if you've chosen a Numerical Flow Structure, such as "10 Reasons to Buy Our Product," then track the countdown (or count up, if you follow David Letterman) for the audience. Let them know where you are at pivotal points in your presentation: "Here's the first reason . . . Now reason number two . . . Finally, here's the tenth and best reason of all . . . "

2. Logical Transition

The Logical Transition is the simplest and most straightforward type of Internal Linkage. It simply means clearly stating the logical connection between one idea and the next.

For example, suppose you are one of a team of presenters offering a detailed account of your company's business and future plans. If you completed your portion of the presentation by saying, "Well, that about wraps up what I have to say. Now I'd like my colleague Nancy to come up and say what she has to say," you wouldn't be making a connection between your material and Nancy's. There would be no logical transition. It would be like a relay racer dropping the baton on the track, forcing the next runner on the team to bend over to pick it up. With this awkward transition, you would let your audience slip from your navigational grasp and allow them to drift. If instead you said, "Now that you've seen the business opportunity and how we're going after it, I'm sure you'd like to know how we're positioned financially to pursue that opportunity. Let me turn the floor over to our CFO, Nancy, who will tell you." That would be a clear linkage. You would be providing *closure of the outbound* (your part of the presentation) and a natural *lead in to the inbound* (Nancy's part of the presentation). You would be passing the baton directly, and retaining your grip on the audience's attention.

Think of the Logical Transition as verbal kin to the visual Bumper slide you read about in Chapter 9, "Using Graphics to Help Your Story Flow." Both transitions provide the closure/lead function, clearing your audience's minds between sections of your presentation in the same way that sorbet cleanses a diner's palate between courses of a fine meal.

3. Cross-Reference

The Cross-Reference is another effective form of linkage. Let's say you introduce a technical concept early in the presentation, but you don't want to drill down with a detailed explanation at that point. Simply make a *forward reference* by saying that you'll cover the concept in greater detail later. Moreover, when you get to that concept, make a *backward reference* by saying, "Now let's turn to the subject I introduced earlier."

Forward and backward references are very powerful tools, if you use them properly. The forward reference, however, can backfire. How many times have you heard a presenter make a forward reference and then fail to deliver? If the audience remembers the reference and realizes that it never materialized, they feel disappointed. Even if they don't consciously remember, they will most likely have a vague feeling that "something was left out." Both alternatives alienate the audience. If you do make a forward reference, remember to complete the circuit by delivering the discussion point you promised.

By contrast, the backward reference is an almost fail-safe tool. When you link back to an idea you presented earlier in your presentation, you get to reinforce that idea. It also indicates that your material is well organized and coherent. The subliminal message: *Effective Management*.

When you are part of a team presentation, you can also cross-reference people. Say one another's names a couple of times during the presentation, preferably first names: "As Frank explained, our R&D division has several of the world's top experts in the field." Such statements send the message that you work well together.

4. Rhetorical Question

For a short, pivotal transition, you can use the Rhetorical Question. Pose a question that grows out of your outbound point or one that leads logically to your inbound point, and then provide the answer.

For example, after describing what your company has done over the past year, you can transition to your future plans with the Rhetorical Question "Where do we go from here?" After explaining how your company intends to take advantage of a major new market opportunity, you can transition to your operating plan by asking, "How will we implement that?" You can transition from your company's business results to a comparison against other companies in the field by asking, "How do we stack up against the competition?" In each case, go on to provide the answer.

Limit your use of the Rhetorical Question; too many of them can sound contrived. The wording can also sound contrived. "Have we set up customer call centers staffed by college-educated personnel to handle complex service issues?" rings artificial. Couch your questions in terms your audience might use, such as "How do we handle complex service issues?"

Limit your use of the Rhetorical Question; too many of them can sound contrived.

---

5. Recurring Theme

A Recurring Theme is a subject that weaves its way throughout your presentation. Let's say you used a customer anecdote about a woman named Louise King as your Opening Gambit. You can then reference Louise King, and the millions of satisfied customers like her, several times throughout your presentation. When you describe the efficiency of your manufacturing processes, you can say, "Because of the low unit cost of our product, we can sell it at a price that people like Louise King can afford." Or, when you explain the marketing campaign that successfully launched your product, you can say, "Louise King saw one of our full-page ads in *USA Today* and called the 800 number listed there."

6. Bookends

An alternative use of the Recurring Theme option is to reference the theme at the beginning of your presentation, and then never mention it again until the end. This form of linkage is called Bookends. You could cite customer Louise

King in your Opening Gambit, and then at the very end say, "Remember Louise King?" The resulting bookend effect provides resolution, and therefore, subliminal satisfaction to your audience.

7. Mantra

A Mantra is a running phrase or slogan that you repeat several times during your presentation. This technique goes all the way back to the Greek orators, who called it *anaphora*.

Many modern-day orators have used the Mantra to great effect. Sir Winston Churchill rallied the beleaguered British people during World War II with a stirring speech in which he repeated the phrase "we shall" 11 times in one paragraph: " . . . we shall go on to the end, we shall fight in France, we shall fight on the seas and oceans, we shall fight with growing confidence and growing strength in the air, we shall defend our Island, whatever the cost may be . . . "

Martin Luther King, Jr., used the phrase "I have a dream" 16 times in his historic civil rights speech.

John F. Kennedy's memorable Inaugural Address is famous for his use of the word "ask" five times in three sentences: "And so, my fellow Americans: *ask* not what your country can do for you; *ask* what you can do for your country. My fellow citizens of the world: *ask* not what America will do for you, but what together we can do for the freedom of man. Finally, whether you are citizens of America or citizens of the world, *ask* of us here the same high standards of strength and sacrifice which we *ask* of you."

What is not as famous is that, within that same 14-minute speech, Kennedy used the word "let" 16 times.

Corporations spend enormous amounts of time and money on specialized marketing consultants to develop corporate slogans or taglines. Think of Microsoft's "Your potential, our passion." or The Human Network's "Are you ready?" or Intel's "Intel inside" or Burger King's "Have it your way."

Your Mantra needn't be just your company's slogan. You can create one that is specific to your presentation. If you were presenting your company's turnaround strategy, and the changes your company has made, you could say repeatedly, "That was then; this is now." If you were trying to persuade new customers to start a trial with your product or service, you could refer to the Chinese proverb that the journey of a thousand miles begins with a single step, and then ask your audience, several times, to take that first step.

The Mantra could just turn out to be the best-remembered phrase or sentence from your presentation, so choose it carefully. If you develop a Mantra, make it pithy, concise, and appealing. Most important, be sure that your Mantra supports your key persuasive theme: your Point B.

8. Internal Summary

The Internal Summary is a way of clearing your audience's minds, and cleansing their palates, by pausing at a pivotal point or two in your presentation and saying, "Let's review what we've covered so far . . . "

Your headlong forward progress is quite well known to you, but brand-new to your audience. They need a moment to digest what you've said and shown. When they don't get that moment, the result can be the dreaded MEGO, or worse, they can interrupt you and throw your presentation off onto a tangent.

No matter how intelligent your audience, they cannot take in your ideas at the rate at which you pour them out. In football, the wide receivers know exactly where they are going, and the defensive backs do not, which gives the receivers a distinct advantage. In football, however, each team gets a chance to switch roles. In presentations, the presenter and audience roles remain fixed, so it is essential that you give your audience the advantage by letting them know exactly where you are and where you are going at all times.

Once you've finished your brief Internal Summary, you can resume your forward progress by leading into the next section of your presentation. Think of this linkage technique as a miniature application of the *Tell 'em what you're gonna tell 'em* concept.

9. Enumeration

Let's say you have four new products in your product line to present. Rather than stepping through the four one at a time, tell your audience that there are four in all. Give them the forest view, rather than one tree at a time.

Introduce them as a group: "We're announcing today a new line of four products, each targeted to a different market segment." Then discuss each one in detail as necessary: "Product A is ideal for the beginner because it offers . . . " After discussing all four, briefly recap the whole product line: "I hope you will see that these four products can serve anyone who needs . . . "

Use Enumeration sparingly, however. Don't give your audience six subthemes under each main topic, and eight subheads under each of the six subthemes. Your audience can't follow your left-brain thinking that far down. They can't keep track of that many levels of detail, nor should you expect them to. When you enumerate, stay with one list, and count your way through it concisely.

10. Do the Math

When you discuss numeric information in your presentation, provide your audience with perspective by comparing, contrasting, or interpreting the numbers for them. For instance, "The debate ran 45 minutes, which means that each candidate had about 22 minutes. In tracking the calls to action, one candidate stated his Point B 21 times and the other 27 times, *an average of one Point B per minute*."

11. Reinforce Point B

As you saw in Story Form, it's important to reinforce your Point B at the beginning and end of your presentation. Experience shows that the two parts of any presentation that audiences remember best are the beginning and the end. Therefore, be sure to highlight your Point B, your call to action, in those key places.

However, in any other than the briefest presentation, you can reinforce your Point B several times. It's the best way to ensure that your audience grasps Point B, remembers it, and understands how each of part of your presentation supports it.

12. Say Your Company Name

When making a business presentation, it's almost unavoidable to refer to your company, so mentioning it is a natural form of linkage. However, be sure to make that reference by name: "Acme Widgets," rather than "our company," or "the company," or "we." This reinforces your company's name in your audience's minds . . . an important consideration since so many presentations occur in industry-wide events, such as conferences or trade shows, where you are competing for mindshare against many other companies.

It is also basic brand identity. Businesses today spend a great deal of time, effort, and money to develop their corporate image in logos, colors, and slogans. They spend even more time, effort, and money disseminating that image on everything from coffee mugs to T-shirts to baseball caps. It is much more time- and cost-efficient to have the company spokesmen and women promote the brand live and in person.

Internal Linkages in Action

How many linkages should you use in your presentation? The best answer is, "Enough." Use as many linkages as you need to create a presentation that is a tightly unified whole. The end result will be a presentation in which each element relates to the next, and all the elements lead to a single conclusion: Point B.

Pick and choose from among the Internal Linkage options that you find appropriate for your presentation and your speaking style. Plan the linkages as you develop your presentation, *practice* them every time you Verbalize, and then deliver them whenever you present.

Plan the linkages as you develop your presentation, practice *them every time you Verbalize, and then deliver them when you present.*

Verbiage

The ultimate phase of bringing your presentation story to life is the words you use to tell your story, or *Verbiage*.

You'll remember from Chapter 4 that the Flow Structure of the Power Presentations program and of this book is Problem/Solution. That is, in each of the essential aspects of a presentation, I show you how *not* to do it, and then how to do it right. To maintain that consistency in addressing verbiage, next you'll find a collection of some of the most commonly used phrases in presentations (you're likely to recognize them, too), each of which will create a problem for you or any presenter. Each problem phrase is followed by a solution, the correct way to state the same idea. The first example is perhaps the most common of all:

▽ **"Now I'd like to . . . "**

Sound familiar? You've probably heard this phrase almost as often as I have. It's virtually boilerplate not only in business presentations, but also in political speeches, college lectures, church sermons, award acceptances, wedding toasts . . . the list is endless.

What's wrong with this phrase? It's *presenter-focused*. It implies that the presenter is making an exclusive decision without any regard for the audience: "I don't care what *you'd* like to do; this is what *I'd* like to do."

It's also vague and indefinite. If you'd like to do it, why not just go ahead and do it? How many times have you been on an airplane and, upon landing, heard the flight attendant say, "I'd like to be the first to welcome you to San Francisco."? How about simply, "Welcome to San Francisco"?

The fix: Make the phrase audience-focused, inclusive, and very definite. Drop the word "like" and simply say:

▲ **"I'm going to talk about . . . "**

Or become even more inclusive by inviting the audience to join you: Shift to the first-person plural using one of these options:

▲ **"Let's look at . . . "**

▲ **"Let's . . . "**

What does the following phrase imply?

▽ **"Like I said . . . "**

This phrase is a form of backward reference, an attempt to link to a point earlier in the presentation. Unfortunately, the specific language suggests that your audience didn't understand your pearls of wisdom the first time you said them, and so you are now going to have to repeat yourself to bring them up to your speed, all of which is condescending to your audience.

The phrase is also poor English. "Like" is the wrong word; it should be "As."

Now this doesn't mean you should avoid backward references. Use them extensively (have you noticed how often you've seen them in this book?). Backward references are powerful tools for continuity and reinforcement.

If you do use them, however, do so with the proper connecting word: "As I said," not "Like I said." But that still leaves the phrase presenter-focused. Better to give your audience credit for having understood and remembered what you said by using one of the following options:

▲ **"As you recall . . . "**

▲ **"We discussed earlier . . . "**

▲ **"You saw . . . "**

▲ **"You'll remember . . . "**

What's the problem with the following phrase?

▽ **"I'll tell you very quickly . . . "**

This phrase implies that you're apologizing for your own material, that what you have to say isn't very important, and so you will hurry through it. By apologizing, you're saying that you didn't care enough about your audience to have prepared your presentation carefully.

There are many variations of the apology, as the following phrases demonstrate:

▽ **"I'm running out of time . . . "**

▽ **"If you could read this slide . . . "**

▽ **"This is a busy slide . . . "**

▽ **"This isn't my slide . . . "**

▽ **"Disregard this . . . "**

▽ **"Before I begin . . . "**

I'll bet you've heard every one of these at least once in your career.

Never apologize, and *always* prepare properly. Omit any topic that does not deserve your audience's time and attention. Present with pride any topic that is important enough to include in your presentation.

What's wrong with this next phrase?

▽ **"Acme listens to its customer and meets his requirements."**

The problem with this phrase is that it's gender-specific. Are all of your customers males? Most likely not, unless your company makes a product like aftershave lotion. The issue here isn't political correctness, but rather accuracy. Make your statement universal by going plural:

▲ **"Acme listens to its customers and meets their requirements."**

In the English language, the plural pronoun "their" has no gender marking, so you can use it to refer to men, women, or both.

Here are three other problematic phrases:

▽ **"We believe . . . "**

▽ **"We think . . . "**

▽ **"We feel . . . "**

Each of these phrases introduces an element of uncertainty. You *believe* something to be true, but is it really? You introduce doubt, even if only subliminally, in the minds of your audience. Your job instead is to convey certainty. The way to get from doubt to certainty is to switch from the conditional to the declarative mood. Recast the entire sentence to eliminate the offending phrase.

Rather than saying:

▽ **"With this large opportunity and our superior technology, I think you'll see that Acme is** positioned for growth."

say:

▲ **"With this large opportunity and our superior technology, you'll see that Acme is positioned for growth."**

The simple removal of the "I think" phrase strengthens the impact of the entire sentence.

This is not to say that, when the outcome is uncertain, you should make forward-looking statements or forecasts. That's risky business. In such cases, use the conditional mood, but instead of using the weak words "think," "believe," and "feel," shift to any of these much stronger options:

▲ **"We're confident . . . "**

▲ **"We're convinced . . . "**

▲ **"We're optimistic . . . "**

▲ **"We expect . . . "**

What's wrong with the following sentence?

▽ **"Acme does not view the competition as significant."**

There are several problems here. For one thing, the statement is arrogant. It suggests that Acme's management isn't taking the issue of competition seriously. For another, it's stupid. Any competition is significant.

On a purely verbal level, however, the sentence is also negative, because of the dreaded "not" word. Most human beings, businesspeople in particular, react negatively to negativity. Recast the sentence to remove the arrogance, the stupidity, and the dreaded "not" word:

▲ **"Acme has strong competitive advantages."**

Here's another form of negativity:

▽ **"What we're not is . . . "**

Far too many presentations begin with a description of the company's business by telling what they are not; instead, tell what *they are*.

For the final problematic phrase, let's flash back to January 1987. President Ronald Reagan had not expressed his position on a very sensitive subject roiling in the media: the involvement of his administration in the Iran-Contra scandal. Reagan finally agreed to tell all in a press conference. The headlines in the next day's newspapers carried his key statement:

▽ **"Mistakes were made."**

But Reagan didn't say *who* made the mistakes.

Now, let's flash forward 10 years. In January 1997, President Bill Clinton had not expressed his position on a very sensitive subject roiling in the media: the involvement of his administration in improper, possibly illegal, fund-raising activities. Clinton finally agreed to tell all in a press conference. The headlines in the next day's newspapers carried his key statement:

▽ **"Mistakes were made."**

History repeats itself. Both Reagan and Clinton couched their statements using *passive voice*. This grammatical construction tells *what* was done, but it doesn't say *who* did it. The result is that the speaker sounds as if he or she is trying to avoid responsibility through fudging and obfuscation.

This is accepted practice in politics, where Reagan and Clinton (and generations of politicians before and after them) were striving to protect their associates, their constituencies, and themselves. This is *not at all* acceptable in business, where *accountability* is paramount. Passive-voice sentences remove the doer of the action, and with it, remove *management* . . . and the presenter . . . from any responsibility or culpability for the action, whether bad or good.

All the previous examples are not meant to be merely a lesson in syntax, but a lesson in psychology. The difference between passive voice and active voice is subtle in grammar, but profound in impact. Avoid the former; use the latter. Put the doer back into the sentence; put *management* back into the equation.

Instead of saying:

▽ **"Mistakes were made."**

▽ **"Progress is being made."**

▽ **"The error rate is being reduced."**

say:

▲ **"We made a mistake."**

▲ **"We're making progress."**

▲ **"We have reduced our error rate."**

When you put the doer back into the action, you put management back into the equation. The subliminal perception is *Effective Management*.

Summary

After you've developed your lucid story and expressive graphics, Verbalize your presentation in advance. Do multiple Verbalizations spaced out over time. As you do, think about and practice the specific language you'll use to convey your ideas. Select and plan clear Internal Linkages. Monitor your Verbiage to develop a positive, respectful, and confident vocabulary.

Just as there are particular styles of slide design that can hinder or enhance your presentations, there are particular words, phrases, and sentences that can offend your audience, or help win them over. By studying your verbal options in advance and by diligently practicing your best choices, you can make your words sing, help your story flow, and bring your presentation to life.

Chapter Eleven

Customizing Your Presentation

Company examples:

- **Elevation Partners**
- **Cisco Systems**

The Power of Customization

On my very first day as an undergraduate at New York University, I attended the orientation session for incoming freshmen in a state of suspended animation. I had left behind the camaraderie and the cinder-block walls of a New York City public high school, and now, alone and apart from my close friends of four years, I entered the marble halls of NYU's Gould Memorial Chapel. I was a solitary speck in a sea of strangers, all of us intimidated by our imposing surroundings and further humbled by our requisite freshmen headdress: bright violet beanies.

My awe was heightened when a procession of austere deans and professors, clad in black academic gowns, entered and took their seats on the stage, looking down at us lowly frosh. As each of these august sages stood in turn to address us, their stentorian tones and crisp articulation echoed off the domed ceiling of the chapel. I was struck by how different they sounded from my high school teachers, whose hardcore New York accents were marked by a strident twang and staccato pace. I was still in the same city, but I might as well have been at Oxford or Cambridge.

Finally, the most impressive and most articulate dean of the group rose to speak. "Gentlemen," he began (for there were no women at NYU at that time), "by now you realize that your life here at the university is going to be vastly different from high school. However, there is one high school practice that we have carried forward. We take attendance."

At that, he reached under his gown, into his coat pocket, and pulled out a set of what looked like business cards. "These are your attendance records," he said. Then, as if holding up a bridge hand, he fanned out the cards in one hand and, with his other, reached forward and plucked one of the cards. Reading from it, he said, "Suppose you're Jerry Weissman . . . "

Imagine my reaction. I couldn't have been more stunned if I'd heard the voice of God intoning my name from the dome.

I wasn't the only one to react. Several other freshmen sitting nearby, whom I'd just met, turned to look at me, and their stirring rippled through the audience like a wave.

I later learned that the dean's selection of my card was completely accidental and arbitrary, but that he always picked an actual card from among the members of each new audience. I also learned that that same dean happened to be the chairman of the Speech Department.

In time, Professor Ormond J. Drake proceeded to become my mentor. Later still, he became my supervisor when I served as an instructor in his department, my sponsor when he gave me a consulting assignment at a Wall Street brokerage house, and my colleague when I produced and directed an episode of his talk-show series for CBS Television. But to this day, what I remember most about Professor Drake is the startling *Aha!* he gave me when he spoke my name in the middle of the welcoming speech he'd given to generations of entering freshmen.

That's the power of Customization.

The Illusion of the First Time

Many business presentations contain information that must be conveyed repeatedly, to multiple audiences. For example, a salesperson may have to present a new product to many different groups of customers, or a human resources manager may have to explain the new company benefits plan to dozens of small groups of employees. In the world of IPO road shows, company officers must make their presentations to many, many groups of investors. Typically, they give 60 to 80 pitches over a period of two to three weeks, often six to eight presentations in any given day.

Under these circumstances, it's difficult to keep your presentation fresh and vital. In part, this is a matter of energy and focus. When you have to make the same points for the third, or tenth, or fiftieth time, it's hard to feel the same sense of enthusiasm, spontaneity, and excitement as the first time. It's all too easy to become bored with your presentation and let your attention flag. When you go into autopilot, however, your presentation comes across as "canned," and the result is an audience that is uninvolved, unmoved, and unconvinced.

The challenge for the presenter is to find ways to overcome this downside, to achieve "the illusion of the first time." This phrase comes from the glossary of stage actors, who often have to perform the same role in the same play hundreds of times (if they're fortunate enough to have a role in a hit production) while conveying to each new audience the sense that every speech and every action is completely spontaneous.

Contrast the theatrical approach to that of the world of journalism. In journalism, an article that can run any time in any edition of a publication is called an "evergreen." This relegates the content of the article to mere filler. *Never, ever* make your presentation an evergreen. As Shakespeare had Hamlet say, "Suit the action to the word, the word to the action." Create the illusion of the first time, every time.

Create the illusion of the first time, every time.

The key to creating that illusion is to make a deliberate effort to focus your energy every time you present. If the thought of repeating the same material over and over again makes your spirit wane, remember the example of baseball immortal Joe DiMaggio. A reporter once said to the Yankee Clipper, "Joe, you always seem to play ball with the same intensity. You run out every grounder and race after every fly ball, even in the dog days of August when the Yankees have a big lead in the pennant race and there's nothing on the line. How do you do it?"

DiMaggio replied, "I always remind myself that there might be someone in the stands who never saw me play before."

In the same way, treat each and every iteration of every presentation of yours as if no one in your audience has ever seen you present before. Make your 80th iteration as fresh as the first . . . and with Verbalization, the first as polished as the 80th.

The equivalent of running out every grounder and racing after every fly ball is to generate your enthusiasm and pump up your energy every time. However, energy alone is not enough. To create the illusion of the first time you must modify your presentation for each new audience. Fortunately, as a business presenter, you enjoy a freedom that stage actors don't have: You can reshape your script and give every performance a new dose of freshness and spontaneity. Does this mean that you have to change your recurring presentation each time? *Not at all.* You can customize the core material with the following techniques; all but one of which involve *only* your narrative. Customize with your words. Use these very same techniques to customize a one-time-only presentation, as well as every presentation you ever give to every audience.

External Linkages

While the Internal Linkages in the previous chapter bond the components of your presentation, it's equally important to bond your presentation (and you, as the presenter) to each specific audience. You can achieve that with External Linkages: words, phrases, stories, and other materials that you insert throughout your presentation to make it fresh. There are seven External Linkages:

1. **Direct Reference.** Mention specifically, by name, one or more members of your audience.
2. **Mutual Reference.** Make reference to a person, company, or organization related to both you and your audience.
3. **Ask Questions.** Address a question directly to one or more members of your audience.
4. **Contemporize.** Make reference to what is happening *today*.
5. **Localize.** Make reference to the venue of your presentation.
6. **Data.** Make reference to current information that links to and supports your message.
7. **Customized Opening Graphic.** Start your presentation with a slide that includes your audience, the location, and the date.

Let's consider each External Linkage, along with illustrative examples.

1. Direct Reference

A Direct Reference is a specific mention, by name, of one or more members of your audience. This is the technique that Professor Ormond Drake used during my orientation at NYU, and its effectiveness is attested to by the fact that I *still* recall the moment, decades later.

There are several ways to incorporate Direct References. One is to refer to audience members to illustrate your key points: "Our services can help reduce the amount of time you spend traveling on business. Take Steve, here, as an example. Steve told me that he's been on the road 12 days this month. With our services, Steve can . . . "

Another way is to tell a story related to the audience or to specific audience members: "As some of you may know, I've worked with your firm before. Last year, Sharon and I developed a joint plan for launching a new program . . . "

"During the break, I was speaking with Howard, and he told me that your company is about to move to new headquarters. You'll be interested to know that our product can help streamline the process . . . "

Be careful to make all Direct References positive and noncontroversial. Only quote statements or tell stories that reveal the audience member in a positive light. And, of course, never violate a confidence.

2. Mutual Reference

A Mutual Reference is a reference to a person, company, or organization that is in some way linked to both you and your audience. Think of a Mutual Reference as a tasteful, appropriate form of name-dropping. For example, in pitching your services to Company A, you might want to describe the work you did for Company B, which has a close business alliance with Company A; or the work you did for Company C, whose CEO sits on the board of Company A; or the work you did for Company D, which is Company A's largest and most respected industry rival.

Before using a Mutual Reference, however, check into the politics of the three-way relationship. Avoid stumbling into a personal or business feud that you didn't know existed. Be certain that your audience will view every connection as an affirmation.

3. Ask Questions

While a question is effective as an Opening Gambit, you can also use the same technique at any point in almost any presentation. By addressing a question directly to one or more members of your audience, you create an effective External Linkage.

There are several ways to use questions. One is the Scott Cook approach of polling the audience. This is a quick way of gauging their interest in or receptivity to a particular concept in your presentation: "How many of your companies plan to increase spending on information technology during the next year? May I have a show of hands? Quite a few, I see. Our new software system can help you get the most out of any new technology you do purchase. Here's how it works . . . "

If you do use this technique, however, be prepared for all contingencies: all of the above, some of the above, and none of the above. Be prepared with a follow-up to each contingency.

Another question technique is to invite audience members to share ideas, reactions, or stories, which you can use as a springboard for further discussion: "Think back to the last time you had a negative experience with airplane travel. What was the problem? Any volunteers? Okay, Reggie, tell us about it . . . "

Yet another technique is to use questions to point the audience toward a predetermined conclusion: "What sort of features would your company need in a new communications system? Where on the list would you rank reliability? . . . And why is that? . . . What happens when your system is down? . . . To address that concern, let's look at some independent data on the reliability of our latest system . . . "

Asking questions is an excellent way of engaging your audience. Getting people to think about issues and discuss them aloud turns your presentation from a one-way transmission into a two-way interaction, increasing your audience's interest and involvement.

Getting people to think about issues and discuss them turns your presentation from a one-way transmission into a two-way interaction.

However, questions do inject a note of unpredictability. It's possible that an audience member's response to one of your questions may raise an irrelevant issue or jump ahead to an idea you will cover later in the presentation. Avoid such detours by politely defining the parameters of your topic.

Preplan the questions you will use, and phrase them carefully. Don't make them heavy-handed and obvious. Based on your knowledge of the audience, word the questions to maximize your chance of eliciting the kinds of answers you want.

4. Contemporize

This technique involves making a reference to the most current of events, what is happening *today*. When you Contemporize, you make it very clear that you have specifically tailored your presentation to your present audience. You send the message that all your information is up-to-the-minute and highly relevant.

Contemporizing is favored by many entertainers, especially stand-up comedians. No monologue by Jay Leno or David Letterman is complete without an assortment of one-liners playing off the day's headlines.

But the technique is equally effective in business presentations. Did the stock market take a sudden nosedive or make a remarkable rally yesterday? Consider linking the news to an explanation of how your company offers its customers increased financial security. Did a local sports team win a major event last night? Consider referencing the event and drawing an analogy to the competitive environment in which your company is operating. In each case, be sure you make the current information link clearly to your main idea.

You can also Contemporize right up to the minute. Refer often to prior speakers, to statements they made, to earlier questions from the audience, or to moments that occurred from the time you entered the presentation environment: the conference room, the auditorium, the office, or even the building. Weave these references throughout your presentation at every opportunity. This technique is the zenith of contemporizing in its immediacy and potency. It is also the easiest to do. All it takes is concentration and memory, and it keeps you fresh every time.

5. Localize

Localizing involves referencing the venue of your presentation. As with contemporizing, it's a favorite among entertainers. Many rock concerts begin with a localized greeting like, "Hello, Philadelphia!," which never fails to draw an appreciative roar in response.

You can Localize your presentation by finding facts about the venue that relate to your message. For example, you can talk about a particular client or customer of yours located in the same city, and then go on to illustrate the benefits your company provided to that client.

You can cite an interesting fact about the city or state that supports your message: "Last year, over 500 patients in this city's hospitals died from drug interactions. Many of those deaths could have been prevented with our automated drug dispensing system."

Or you can refer to a noteworthy local person, landmark, or incident, drawing a connection to your offering:

"It's good to be here in St. Louis, where one of America's favorite treats was invented over a hundred years ago. An ice cream vendor at the St. Louis World's Fair ran out of paper cups on a sweltering afternoon. Desperate for a way to keep serving his customers, he got together with the waffle vendor next door . . . and the ice cream cone was born. Today, we're presenting a new product that embodies the same kind of entrepreneurial creativity . . . "

6. Data

You can also create an External Linkage by citing current Data that links to and supports your message. The more up-to-the-minute and closely linked the Data is to your specific audience, the better. If the Data you mention is news to your audience, they will be impressed by the depth and currency of your information. If your audience members are already aware of the Data, they will be quietly pleased that you are as knowledgeable as they are. Either way, you create a positive link.

To add a touch of emphasis, use the source of your Data as a prop: "Have you seen today's *The Wall Street Journal*?" (Hold it up.) "There's a striking graph on the front page that shows just how serious our industry's infrastructure problems have become." (Quote the most relevant number.) "This is exactly the issue our new system has been designed to address."

7. Customized Opening Graphic

The final and simplest type of External Linkage is the Customized Opening Graphic. Begin the visual portion of your presentation with a slide that shows the audience, the location, and the date of your presentation. At the start of every Power Presentations program, I use such a slide. In fact, I also include the logo of the company I am coaching.

A Customized Opening Graphic may seem like a small item, but it has a powerful effect on both the presenter and the audience. For the presenter, it forces a final double-check to avoid the embarrassment of the wrong slide. It's also an up-front prompt that gives an impetus to the style and content of the rest of the presentation.

For the audience, it sends the message that you've prepared this presentation especially for *them* . . . that it is not a generic recitation, but rather a custom-made work, tailored to their needs and interests. Thus, the slide launches the presentation with a clear stroke of Audience Advocacy, promising in effect that you are there to serve them.

Roger McNamee, a managing director and cofounder of Elevation Partners, is one of the most influential investors in the technology sector, and one of the best presenters I know. He is always in great demand to offer his unique views at major industry conferences. Roger carefully crafts each speech in advance and is keenly aware of the importance of Customization. Here's how he puts it:

> *Speaking engagements are great branding opportunities, but only if you do a good job. You can do enormous harm to yourself by not understanding the opportunity. It's a crime to give the "windup doll" speech, one that should be punishable by more than just not being invited back.*
>
> *If you don't feel some level of affinity for the community you are addressing, don't do it. If you do, you will only lessen the value of your brand. A speech is like a shark. It can't sit still; it either is building or destroying your brand.*
>
> *Personal brands have a life of their own, dependent on the perceptions of others. It's like cartoon characters who run off the edge of a cliff: They keep running in space until they realize there is nothing supporting them, at which point they plummet. A brand is the same: It's at the moment when everyone comes to believe that there is nothing supporting you that your brand collapses. Public perception is everything.*
>
> *It's death if you give a speech or a presentation that sounds like a prerecorded announcement. Odds are that you don't understand the critical success factors in your business well enough if you don't understand this truth.*

Roger heeds his own advice by starting each speech from scratch, but his words are applicable to *every* iteration of *every* presentation.

Gathering Material for Customization

To customize your presentation, arm yourself with useful information and materials. Begin this process during your preparation period, days or even weeks before the presentation, and continue right up until the moment you approach the front of the room. Here are some steps you can take:

Prior to Presentation Day

- **Research your audience.** Learn all you can about who will be attending: their knowledge level, key interests and concerns, and personal or professional biases.
- **Learn the names of some key audience members.** Know the names of several key influencers in the audience. Learn the names of the highest-ranking company officer, the most respected technical expert, and the manager with the most authority to make decisions.
- **Get current on industry news and trends.** During the run-up to your presentation, diligently search out news and media stories and Internet items related to the company and the industry to which you'll be presenting.

On the Day of the Presentation

- **Customize your Opening Graphic.** Produce an initial slide that names the audience, venue, and date of your presentation, and tee it up to launch your program. Microsoft PowerPoint has a specific function that changes the date automatically. (On the toolbar, click Insert, and then click Date and Time, and then click Update automatically.)
- **Search for ways to Contemporize your presentation.** When you awaken on the day of your presentation, watch the morning business channels on television, read the daily newspapers, and log onto the Internet; browse all these sources to find items relevant to your presentation and your audience.

  The New York Times runs a daily feature called "This Date in Baseball" with memorable milestone events. Since competitive sports are an excellent metaphor for business, you could choose one of the events to analogize and illustrate your situation.

  On a broader scale, a website (www.scopesys.com/today/) lists significant events on any given date in history. On the day you present, find an event that parallels your story, and incorporate it to add dimension to your presentation.
- **Prior to your presentation, mingle with your audience.** Go into your audience and chat with several individuals (this is known as "schmoozing"). Choose strangers as well as people you know. Ask them questions. Listen to their conversations. Gather valuable information as well as names and facts you can incorporate into your presentation.

Don Listwin, now the founder of The Canary Foundation, a nonprofit organization promoting early detection of cancer, was for many years the Executive Vice President of Cisco Systems. I first met Don when he was a junior product manager at Cisco. You'll recall from the Introduction that after Cisco's IPO, Cate Muther, then the Vice President of Corporate Marketing, required that all her product managers take my program. Don was in the first wave, and he was a most diligent student. Before long, his advanced presentation skills earned him a plum assignment: He was chosen to announce the launch of a major new Cisco product alongside then-CEO John Morgridge.

This was a highly mission-critical task. Since its inception, Cisco had been engaged in a fierce competitive battle with Wellfleet for leadership of the router market. In 1992, Cisco produced a new integrated router that would strike directly at the heart of Wellfleet's strength, hardware, and exploit the vulnerable underbelly of Wellfleet's weakness, software, which happened to be Cisco's special strength.

Don crafted his presentation for the media using all the techniques in the Power Presentations program, and then spent the weekend before the scheduled Tuesday launch Verbalizing in front of a mirror *40* times. On Monday, Don did a trial run in front of an internal Cisco audience, and went into autopilot. His presentation had become dry and robotic.

He called me and said, "Jerry, I've gone stale! What do I do?"

I reminded Don about the Customization techniques, and he seized upon Direct References and Questions. The next day, just before his presentation, Don went into the audience and chatted with several people, asking them what they were hoping to hear from Cisco. Then he stepped up to the stage and began his presentation. Right after his Opening Gambit, Don looked at one of the people with whom he'd chatted and addressed him by name. The man smiled, and Don felt the spark of recognition radiate through the crowd. In turn, Don felt in control of himself and his audience. (Later, Don likened this effect to what happens when a student is called on by a professor, although I had never told him the story of Professor Drake.)

Then Don raised one of the questions someone had asked him during his schmooze, "Will Cisco continue to upgrade the performance of this new router?" Don promptly answered it by describing how Cisco planned to migrate the new product forward. Once again, Don could feel the energy from the audience, and it energized him in turn. He rolled forward with a full head of steam.

Less than a year later, Cisco's new router product was the clear market leader, and a year after that, Wellfleet disappeared in a merger.

Several years later, I was delivering my program to a group of new product managers at Cisco. When I got to Customization, I referenced Don's example. One of the men in the group exclaimed, "Oh yes! I was with Wellfleet at the time, and I was in the audience for that *prezo*." He shook his head ruefully. "After I heard Don speak, I knew that the game was over."

Don still practices these techniques to this day to keep his presentations fresh and to connect with his audiences.

External Linkages in Action

Customizing your presentation is an art, and, as with any art, it takes practice to perfect. Try your hand at Customization the next time you present, and practice it every time you make a presentation in the future. The time you invest in making each presentation unique will pay handsome dividends in the form of greater audience involvement, and many more priceless *Aha!*s.

The same techniques that make your presentation more timely, relevant, and compelling to your audience can also make preparing and delivering it more creative, spontaneous, and stimulating for you. The entire process serves as a feedback loop to invigorate you with the same kind of hustle Joe DiMaggio exhibited.

I practice what I preach. I've been delivering the same core material to my clients for 20 years. If I were to deliver it the same way every time, I'd be on full autopilot by now. Instead, I simply open the air vents of my mind, take in data about my audience, and then circulate it back out to them. It makes them feel involved, and it makes me feel energized.

While many of the External Linkages require considerable but worthwhile prior effort, the most effective Customization requires no preparation at all. It all takes place "live and in person," and is the ultimate form of Audience Advocacy: Concentrate on your audience *during* the presentation. This means weaving in many Direct References to members of your audience and making many Contemporizing references to moments that have occurred since your presentation began. All it takes is concentration: Be in the moment. When you are in the moment, you will feel a connection with your audience, just as Don Listwin did.

Concentrate on your audience during *the presentation. Be in the moment.*

Of all the many presentation methods I've given my clients over the years, the Customization techniques are implemented the least. Yet they provide potentially the biggest bang for the buck. They provide the most powerful ways to differentiate your presentations from the routine, impersonal, one-size-fits-all, plug-and-play presentations we see all too often. Learn these techniques, and implement them the most.

Professor Drake's lesson inspired me and has stayed with me all this time. Let it become the big bang for *your* buck.

Chapter Twelve

Animating Your Graphics

COMPANY EXAMPLE:

- **Microsoft PowerPoint**

How Versus Why and Wherefore

What comes to mind when you see the word "animation"? Maybe a vintage Bugs Bunny short or a more recent full-length feature, such as DreamWorks' *Shrek* or Pixar's *Ratatouille or WALL-E*. In the context of business presentations, animation refers to motion added to computer graphics. This movement can involve either an entire slide or the visual elements within a slide. In animation, these elements move onto or off the screen; or shift within the screen; or grow, shrink, change, or vanish.

We've all seen varying degrees of electronic animation in business, from the sophisticated sequences that appear on websites to the equally sophisticated presentations at industry conferences and trade shows, many of them worthy of Disney or Pixar. Often, even conventional prepackaged corporate pitches have screen effects that rival the production values of the big-tent special events.

Most of these examples of animation are created by professional graphic artists and technicians using complex software, such as Adobe Director or Flash. Professional artists also use Adobe Photoshop to render objects and images in vivid, opulent detail for animation as well as for conventional presentations.

For the rest of us, the vast universe of consumers, there is Microsoft PowerPoint, which is installed on hundreds of millions of computers that churn out 30 million presentations per day.

This ubiquitous software, launched in 1987, has been growing its market share with each successive release, and now it is standard operating procedure for business presentations. How often have you been asked to send someone a copy of your PowerPoint slides? How often have you sat next to someone on an airplane clicking through his or her slide show? PowerPoint has also reached beyond business into our daily lives. Even elementary schoolchildren use it expertly.

Yet, as a presentations coach, one of the most frequent complaints my business clients make is, "I'm not a graphic artist!" As a result, they default to the Presentation-as-Document Syndrome. But, as you discovered in Chapters 6 through 9, by simply going beyond text to pictorial, numeric, or relational images and applying the principles of *Less Is More* and *Minimize Eye Sweeps* to those images, you can create graphics that are both interesting and effective.

Creating animation for those graphics, however, can be daunting. We've all been in the audiences of far too many presentations that unleash all the bells and whistles of the animation in PowerPoint with a frenetic, pyrotechnic display that challenges a Fourth of July celebration. This phenomenon is like putting a 14-year-old boy behind the wheel of a Ferrari Testarossa.

This situation reached epidemic proportions in the military. Pentagon officers were launching so many PowerPoint-powered animated tanks and whirling pie charts in their slide shows that the chairman of the Joint Chiefs of Staff was driven to issue a directive demanding simplicity in Defense Department presentations.

It would seem that military personnel, as well as businesspeople . . . basically conservative segments of the population by nature . . . should be aware that flashy animation projects a negative image. And yet, business presentations that look like MTV videos are as common as car crashes in Hollywood movies . . . and just as irritating.

The reason that businesspeople perpetrate this graphic assault on their audience's visual senses is because these presenters have learned the *how-to* of animation, but not the *why* and *wherefore* of its application. The computer sections of bookstores abound with worthy PowerPoint books. The Web, television, newspapers, and magazines abound with equally worthy courses and courseware that provide excellent instruction about how to operate and navigate the software. None of them, however, tells you *why*, *where*, and *when* to use animation, and particularly, not what effect animation should or could achieve. You'll find those answers in this chapter.

Presenters have learned the how-to *of animation,*
but not the why *and* wherefore *of its application.*

Why use animation at all? As a conservative businessperson, you might think that animation is irrelevant, frivolous, or unnecessary. "I'm not up there to entertain people," you might say. "When I'm making a presentation, people just want the facts, plain and simple. Fancy gimmicks will just take away from my message."

You may be right. Any visual aid can indeed become a visual hindrance when it's misused, resulting in distraction, annoyance, or confusion in your audience. So it is with animation. But any sword can cut both ways.

The word "animation" comes from the Latin root *anima*, which means "spirit" or "life," just as the word "animated" describes a lively or energetic person. Animating the graphics in your presentation can add a sense of spirit and life to what might otherwise be a flat visual display.

Even more important, well-designed and appropriately applied animation can actually enhance your message. Just as you can create text, pictorial, numeric, and relational slides to express your important concepts, you can also strengthen that expression by adding animation to bring graphic objects on or off the screen meaningfully. The right animation can make your presentation more visually appealing, transforming it from the merely good to the truly captivating . . . and therefore persuasive.

The right animation can make your presentation more visually appealing,
transforming it from the merely good to the truly captivating . . .
and therefore persuasive.

The answers to the *wherefore* of animation can be found in intrinsic human perception and in cinema, the same sources that provided the fundamentals for the design of presentation graphics that you saw in the earlier chapters. The core principles of cinematography and editing are even more relevant in animation. What follows is an extrapolation of those well-known, innate, and well-established professional principles into a simple set of guidelines that you can apply to animate your PowerPoint presentations. We begin with the human factor.

Perception Psychology

The operative rule for designing animation effects goes back to Ludwig Mies Van der Rohe's *Less Is More* principle. Simplicity is the watchword for the graphics in any presentation, and that applies to animation as well. Moreover, whenever motion is involved, we must also keep in mind the cultural, psychological, and neurological factors that influence how people perceive and process visual cues.

In Chapter 6, "Communicating Visually," you read that psychologist and art historian Rudolf Arnheim, in his book *Art and Visual Perception: A Psychology of the Creative Eye*, described the tendency of the human eye to move across a visual field from left to right. This innate effect is further heightened in Western cultures. Text in Western languages (including English) is printed from left to right. These predispositions have a profound impact on how *all* human beings . . . including presentation audiences . . . perceive visual stimuli. Whenever our eyes move from left to right, the information we absorb feels "natural," "normal," "smooth," "easy," and "positive." Many of the visual arts follow this same path:

- On the stage, protagonists usually move to the right (sympathetic movement), and villains move to the left (asympathetic).
- In the cinema, a pan right is positive and fluid; a pan left is negative and drags.
- In heraldry (the design of coats of arms), a crest that has a diagonal bar slanting down to the right is known as a *Bar Dexter* (from the Latin *dexter*, meaning right) and is said to represent legitimate members of a family. A crest that has a diagonal bar slanting down to the left is known as a *Bar Sinister* (from the Latin *sinister*, meaning left) and is said to represent a bastard. Although heraldic scholars debate the validity of these meanings, by looking at Figure 12.1, you can readily see that the *Bar Dexter* flows easily, whereas the *Bar Sinister* drags backwards.
- Even language echoes our innate preference for the right side: *dexterous* means skillful or capable, whereas *sinister* means evil or malevolent.

Figure 12.1 *Bar Dexter versus Bar Sinister.*

For these reasons, if you want your presentation audience to feel positive about your ideas, your animation should follow the natural, reflexive eye movement: left to right. Of course, if you want to send a negative message . . . say, about your competition . . . you should reverse direction, and move your objects right to left. But do it deliberately; don't send mixed signals by delivering a positive message about you or your business by making a negative move. As you will see in this chapter, motion can induce a variety of other psychological perceptions. Because the greater part of any presentation is about you or your company, send positive messages; make the default direction of your animation left to right.

Send positive messages;
make the default direction of your animation left to right.

In addition to these innate emotional factors, your audience's eyes are also driven by their highly light-sensitive optic nerves. When motion occurs on the presentation screen, the audience looks at the moving image *involuntarily*. If that movement is counter to the message you are trying to convey, you will confuse the audience for an instant. Such instants can build into a giant MEGO at best, or complete resistance to your ideas at worst. If the movement supports your message, your audience will stay with you . . . and be more receptive to your ideas.

Cinematic Techniques

In cinema, directors and editors use the camera and a montage of camera shots to express the emotional qualities of a story. The movement of subjects in front of the camera and the movement of the camera itself, along with the juxtaposition of the shots, can create either positive or negative feelings. In a romantic movie, when long-separated lovers finally come together in an embrace, the scene is likely to be filmed in long, smooth, flowing shots, conveying sensuality and abandon. In a cops-and-robbers drama, a car chase is usually captured in sharp angles and edited with short, rapid cuts, creating tension. In a western, when a wagon train of settlers moves across the screen, the camera slowly draws back from a close-up until the entire panorama of the prairie is visible, expressing the vast challenge of their journey. And in a suspenseful murder mystery, when the detective, searching for clues in a dark, empty house, suddenly hears a click, the camera quickly cuts to a close-up of a door handle, heightening the tension.

Presentation animation doesn't offer the range and power of options available to a film director, but all of the techniques you've just learned can be distilled into one a simple overarching principle: *Use motion to help tell your story by expressing the action in your message*; use motion to mirror or evoke the feeling you want to create in your audience. To reprise Shakespeare: "Suit the action to the word, the word to the action."

Use motion to help tell your story by expressing the action in your message.
Use motion to mirror or evoke the feeling
you want to create in your audience.

The tools to implement this cinematic approach exist in abundance in PowerPoint.

Microsoft PowerPoint 2003 and 2007

PowerPoint didn't start out as fully developed as other applications in the Microsoft Office suite, but as the various generations of the program have evolved over the years, Microsoft has continually added new and improved capabilities. The latest version, PowerPoint 2007, features a complete overhaul of the program's interface, adopting Microsoft's new "ribbon" interface in place of the traditional pull-down menus.

However, most of the features found on PowerPoint 2007 ribbons can also be found on the pull-down menus of previous versions. This means that just about everything you can do with PowerPoint 2007, you can also do with previous versions, such as the still widely used PowerPoint 2003.

Whichever version of PowerPoint you're using, there are two primary forms of animation: Slide Transition and Custom Animation. Slide Transition creates the effect of movement from slide *to* slide (interslide); Custom Animation creates the effect of movement of text and graphic objects *within* a slide (intraslide).

Custom Animation has four major categories of transitions: Entrance (moving an item onto the slide), Exit (moving an item off the slide), Emphasis (making an item more prominent on the slide), and Motion Paths (moving an item across the slide). When you choose a transition type, you also choose the type of transition effect you want, such as Blinds, Box, or Fly In. For each effect, PowerPoint offers multiple directional choices (Up, Down, Left, Right, In, Out) as well as multiple speeds (Very Fast, Fast, Medium, Slow, Very Slow).

Similar options exist for PowerPoint's Slide Transition effects. You can choose from 58 different effects grouped into five categories: Fades and Dissolves, Wipes, Push and Cover, Stripes and Bars, and Random. You also get three speed options for each effect: Slow, Medium, or Fast. You can apply the same effect to all the slides in your presentation, or vary the transition effects from slide to slide.

Additionally, 19 predefined sound effects are available to accompany the graphic animation; you can even add your own sound files as audio effects. But because the focus of this chapter is *visual* animation, all I'll say about sound is to invoke again Ludwig Mies van der Rohe's surgically appropriate advice: *Less Is More*.

This broad array of graphic options constitutes a rich repertory of tools you can employ to animate your presentation. Take the time to surf through the menus and see how the effects play out. Your first reaction to many of them will be, "Wow! Cool!" Your second reaction may be, "But when would I ever use *that*?" The more extreme a particular effect appears, the more specialized the circumstances in which it might be appropriate. In the next section, we'll discuss the whys and wherefores of some of the most frequently used effects . . . as well as some of the extreme ones. Learning to use these new tools is like learning to use the features of a new digital camera or mobile phone: It takes time, learning, and practice. Experiment with them. The results will be worth your while. Once you gain mastery, you'll be able to take *any* presentation from the static to the dynamic.

Animation Options

Now let's see how you can implement Perception Psychology and cinematic techniques in your presentation with PowerPoint. The best place to begin is to restate the overarching idea drawn from the cinema: *Use motion to help tell your story by expressing the action in your message.* Use animation to make a point, disclose content, connect ideas, or create conflict. Gratuitous animation distracts your audience. To keep this discussion at the macro level, we will focus on only a few of the many options and the actions they describe. The options happen to be the ones I use most frequently in my programs and seminars. I practice what I preach.

Use motion to help tell your story by expressing the action in your message.

Please note that most of these choices exist in both Slide Transition and Custom Animation. Also note that when you animate individual text elements or objects on a slide, many of the choices are available for both item Entrance and Exit. These variations provide enormous flexibility to create dynamic animation.

To begin, open a PowerPoint presentation and select a text box or graphic. In PowerPoint 2003, you go to the top toolbar and click Slide Show. When the pull-down menu appears, click Side Transition to see the Slide Transition animations, or click Custom Animation to open the full animation menu.

In PowerPoint 2007, select the Animations ribbon, shown in Figure 12.2. This ribbon displays the 58 Slide Transition animations; to apply a transition effect to the current slide, simply click the transition you want. You can then pull down the Transition Speed list to select the speed of the transition. And, if you want to apply this transition to all the slides in your presentation, click the Apply to All option.

Now let's examine the animation effects you can apply to individual objects or blocks of text on a slide. Begin by selecting the object or text block you want to animate, and then either select Slide Show, Custom Animation (PowerPoint 2003) or click the Custom Animation button on the Animations ribbon (PowerPoint 2007). With either version of the program, this displays the Custom

Animation pane on the right side of the PowerPoint window. From here, click the Add Effect button, and then select the type of effect you want to add. For example, if you click the Add Effect button, click Entrance, and then click Wipe, you'll see four Wipe choices: From Bottom, From Left, From Right, and From Top. Select Wipe From Left. This is the ideal animation choice, because it flows with the natural, reflexive, left-to-right movement that our eyes find so pleasing. Consider this choice your default for all text.

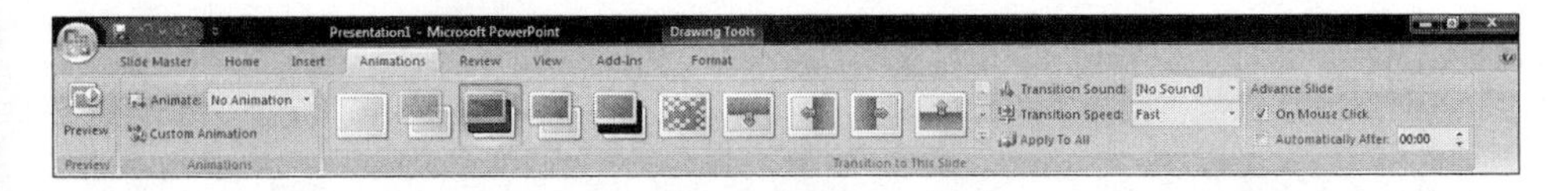

Figure 12.2 *Select various animation effects on PowerPoint 2007's Animations ribbon.*

This is the ideal animation choice, because it flows with the natural, reflexive, left-to-right movement that our eyes find so pleasing. Consider this choice your default for all text.

Wipe From Left/Wipe Right

The Wipe From Left choice in Custom Animation produces the same effect as the Wipe Right choice in Slide Transition . . . both wipe to the right. For the purposes of this discussion, let's refer to both as Wipe Right. For many presenters, Wipe Right is the only animation choice they'll ever need. This option is available for both slide transitions and text/object animations. Unless you want to send a different visual message, as described in the following section, make Wipe Right your default choice for every click in every presentation.

This Wipe Right option can be further reinforced by always using the title of the Slide Master as an Anchor Object. To display the Slide Master (see Figure 12.3), which controls the default formatting of all new slides in a presentation, select View, Master, Slide Master (PowerPoint 2003), or select the View ribbon and click Slide Master (PowerPoint 2007). With the Slide Master displayed, left-justify the Master Title text, and then create the title of every slide using this default. Be sure to always use the same size of font for each title, and, of course, keep each title to a single line (*Minimize Eye Sweeps*). With this configuration, when you run your slide show, the title of each new slide will replace the title of the outgoing slide in precisely the same place. The Anchor Object animated.

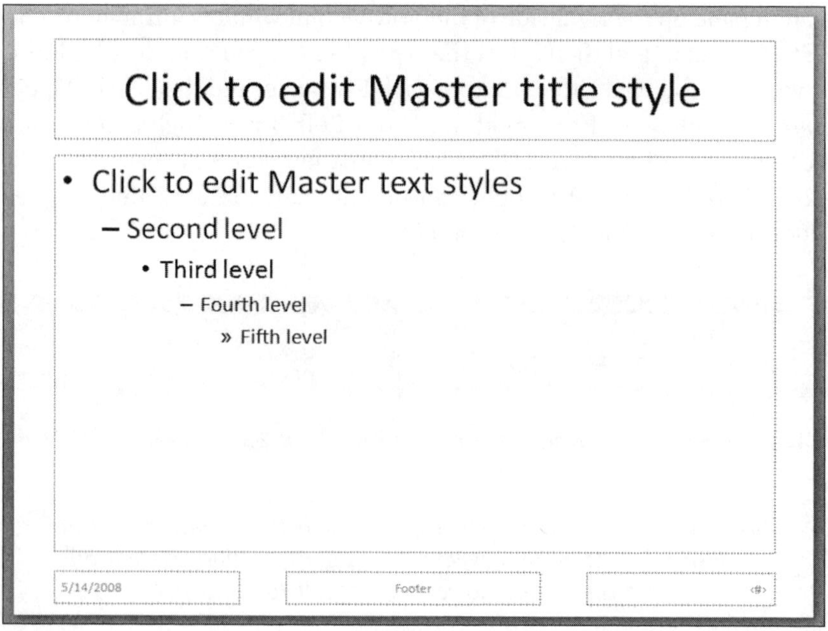

Figure 12.3 *The Slide Master.*

Remember the Conditioned Carriage Return from Chapter 6? Your audience is culturally accustomed to start each new slide at the upper-left corner just as they do in a book. The Wipe Right animation of the title will create the familiar . . . and positive . . . impression of turning pages in a book.

Wipe Left/Wipe From Right

The Wipe Left option, the opposite of the Wipe Right effect, is also available as both a Slide Transition and a Custom Animation (where it is called Wipe From Right). It produces the effect of movement *against* the grain. Use this effect only if you want to send a negative message to your audience. The entrance of *any* graphic object from the right, particularly text, forces your audience's eyes to drag backwards. I use the Wipe Left in the slides in my programs to indicate what *not* to do. The subtle sense of discomfort my clients feel in watching the counter movement of images reinforces the point of my message.

You can use the Wipe Left in your presentations to indicate the negative factors in your story: the shortcomings of competing products, past problems your company has conquered, or market forces that pose major obstacles for your industry.

Fly and Peek

The Fly and Peek Custom Animation effects are seemingly close cousins of the Wipe, but beware: Although they bring text and objects onto the screen, as the Wipe does, the Fly and Peek versions blur the text or the object as it moves across the screen, making the image unreadable. Of course, if you want to suggest a feeling of speed or haste, use Fly or Peek. The blurring is more pronounced with text than with pictures, so even if you want to express haste, convert your text and data into pictorial images . . . a good rule to follow for *all* slides. If you want your audience to take in your images comfortably or read your text easily, stay with the Wipe.

No Transition, Cut, Blinds, Checkerboard, Comb, Random Bars, and Newsflash

This group of Slide Transitions, like the Fly and Peek Custom Animation effects, also tends to jar the audience. The No Transition and Cut options pop new information onto the screen abruptly. These are PowerPoint versions of what is known in the cinema as the crash cut. Filmmakers use this device to create sudden juxtapositions. Use the No Transition or Cut option in your presentation if you want to express sharp contrast between two slides, such as the difference between the sleek new design of your logo and the busy and cluttered logo of your chief competitor.

The Blinds, Checkerboard, Comb, Random Bars, and Newsflash options are all very dramatic and call attention to themselves. Used arbitrarily, they could distract from your presentation, but *to everything there is a season, and a time to every purpose under the sun*. The Blinds, Comb, and Random Bars do make effective transitions between bar charts, and the Checkerboard does create an effective segue between tables. You'll recognize the Newsflash effect from vintage Hollywood movies, as when the front page of a newspaper comes spiraling onto the screen. Although theatrical, you can use it in a presentation to introduce a page from a publication highlighting a favorable quotation endorsing your business, product, or service.

Use any of these flamboyant options with care. Be aware of the potential downside: Any animation that provokes an audience response of "Huh?" or "What was that?" or "I don't get it!" only detracts from your message. When in doubt, default to the Wipe Right.

Cover, Push, Strips, and Uncover

The Cover, Push, Strips, and Uncover Slide Transitions are effective variations of the Wipe in that they all introduce new images with significant but less extreme visual movement. However, keep in mind the basic rule that the direction of the motion should express the action in your message. For instance, if you bring in a new slide using the Cover Down option, it has the finality of a curtain descending . . . hardly the kind of message you want to send about your business. The Uncover Down effect, however, evokes the sense of unveiling a work of art.

Keep in mind the basic rule that the direction of the motion should express the action.

Dissolve and Fade

The Dissolve and Fade options provide smooth, as well as tried-and-true, choices for transitions. They are familiar to everyone from their antecedents in cinema, where they are used to indicate the passage of time. In films, an outbound image gradually fades out as the inbound image gradually emerges. The images overlap in the Dissolve and dip to black for an instant in the Fade. You've seen these effects countless times to indicate the shift of night into day, of one season into the next, and of youth into old age.

In PowerPoint, the Dissolve and Fade are available as Slide Transitions and Custom Animation (as both Entrance and Exit effects). Consider using a Dissolve in a presentation to indicate a segue between related concepts. For example, an image of a product could dissolve into an image of the new, upgraded version of the same product. In contrast, an image of a metastasized cell could shrink or disappear, thanks to the power of a new drug. Beware of Fades, however. When images Fade In or Out, they create a sense of detachment or separation. Business presentations, however, are not detached; they deal with closely related subjects in real time. (Think about that: today's business presentations observe the classic Aristotelian unities!) Therefore, use the Fade very, very sparingly. Moreover, Fades and Dissolves, at any speed, appear slower than the preferable Wipe and tend to drag the pace of a presentation. Choose animations that express continuity and connectivity. Remember: Grab, Navigate . . . *never* let go . . . and Deposit at Point B!

Wipe Up/Wipe From Bottom and Wipe Down/Wipe From Top

Here again, PowerPoint uses different nomenclature in Slide Transition and Custom Animation for the same effect. And here again, for the purposes of discussion, we will refer to all of them as Wipe Up or Down. In these choices, the motion has a direct relationship to the action. Use Wipe Up to indicate rising revenues, increasing market share, or growing profits. Use Wipe Down to indicate declining expenses, a falling error rate, or reduced employee turnover.

Circle, Box, and Split

The Circle, Box, and Split options express either compression or expansion, depending on whether their movement is In or Out. Choose the Custom Animation that relates to the innate shape of the compressing or expanding object. Here are a few ideas:

- The **Circle Out** effect can express a pie chart or globe that grows, or a hub expanding out to its spoke structure. The Circle In effect can express the shrinking or condensing of a pie chart or globe. The direction of your animation depends on the message you want to deliver.
- The **Box** effect can express a grid or table that narrows or widens.
- A **Split In** can represent a multiplicity of diverse objects coalescing into a single whole.
- A **Split Out** can represent a central theme with multiple variations, such as a central server with many clients or a central office with multiple branches. You can also use the Split Out to create the effect of opening a curtain.

Wheel and Wedge

The Wheel and Wedge Slide Transitions are effective ways of replacing one image with another. Reminiscent of the iris effect from silent films, the Wheel and Wedge use a circular motion opening out or closing in to create a transition.

One particularly effective use of the Wheel option is to highlight a single word, number, or object on a slide in the same manner as you would draw a highlighting circle around a section of a document. You use PowerPoint to animate the circle by using the Wheel effect as follows:

In PowerPoint 2003, open the Drawing toolbar and create your circle or oval. Then, with the new object selected, continue along the Drawing toolbar to Fill Color, and select No Fill. Next, move to Line Color and select a bright or contrasting color. Then move to Line Style and select 6 point.

In PowerPoint 2007, the process is slightly different. Select the Insert ribbon, click the Shapes button, and create a circle or oval. Double-click the circle or oval to display the Format ribbon, and then pull down the Shape Fill list and select No Fill. Next, pull down the Shape Outline list and select a bright or contrasting color; still in this pull-down list, click Weight and select 6 point.

With this formatting completed, position the circle or oval over the word or number in your graphic you want to highlight, as shown in Figure 12.4. Now you are ready to animate your highlighting option.

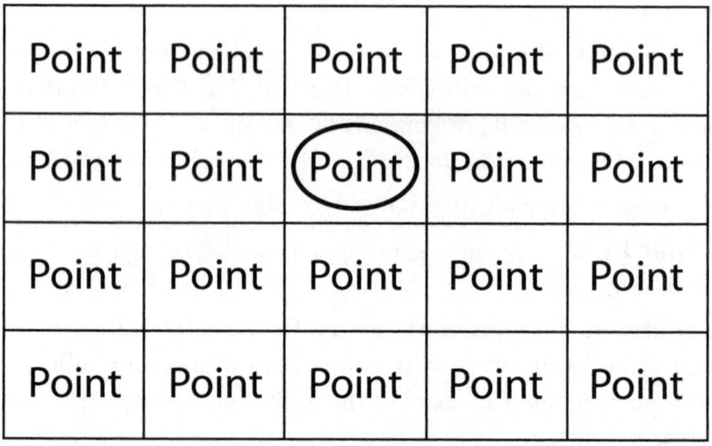

Point	Point	Point	Point	Point
Point	Point	Point	Point	Point
Point	Point	Point	Point	Point
Point	Point	Point	Point	Point

Figure 12.4 *An oval highlighting one word, positioned for Wheel animation.*

With the circle or oval selected, in PowerPoint 2003 select Slide Show, Custom Animation. In PowerPoint 2007, select the Animations ribbon and then click Custom Animation to display the Custom Animation pane. Then, in both versions, click Add Effect, click Entrance, click More Effects, under Basic click Wheel, and then click OK. In the Spokes window, select 1, and in the Speed window, select Very Fast. Check the action in Slide Show mode.

You can use a similar approach to animate a U-turn Arrow. An often-needed tactic in business is a change of direction. In PowerPoint 2003, select the U-turn Arrow from AutoShapes in the Drawing toolbar. In PowerPoint 2007, you can express this action in a diagram or flowchart by selecting the Insert ribbon, clicking Shapes, and then selecting the U-turn Arrow object (in the Block Arrows section). You can then introduce the figure using the Wedge animation in the same manner as the Wheel example.

Emphasis

Another way to call attention to a word, number, or object on a slide is to use the Custom Animation Emphasis effects to Change Font, Change Font Style, Change Font Size, Grow/Shrink, or Spin the word or object you want to accent. (Other emphasis effects are also available by clicking More Effects.) This feature provides multiple ways to implement such highlighting. There are so many effects and colors available with this feature, it is best to browse through all of them to see which works best for you.

One of the most powerful tools on the Emphasis menu is the Grow/Shrink option. With this choice, you can enlarge or reduce any object, such as an oval or a rectangle with text inside. This will allow you to call out the object and discuss it in detail.

To do this in PowerPoint 2003, select Slide Show, Custom Animation. In PowerPoint 2007, select your object, select the Animations ribbon, and click Custom Animation to display the Custom Animation pane. Then click Add Effect, click Emphasis, and click Grow/Shrink. In the Size box, select a number greater than 100 percent, and press Enter. The object will grow. To shrink the enlarged object back to its original size, select your object again, click Add Effect, click Emphasis, click Grow/Shrink, and enter a number smaller than 100 percent in the Size box.

This technique is an advanced extension of the Anchor Objects you saw in Chapter 9, "Using Graphics to Help Your Story Flow." All the objects remain in constant view of your audience as you work your way through the entire slide, enlarging specific objects one at a time. The audience never loses sight of the big picture as you *navigate* them through your presentation, discussing and adding value as you go.

Motion Paths

Motion Paths is the most sophisticated of PowerPoint's animation techniques. It is the one feature that comes closest to the animation capabilities available in professional television control rooms and in computer-generated animation cinema studios. With Motion Paths, you can move objects from one part of the screen to another without having to lose the object or exit the slide. This enables the motion to make all kinds of expressive statements: escalating revenue potential, declining costs, shifting strategic directions, realigning alliances, or rotating positions. To achieve these, select an object and display the Custom Animation pane. Then click Add Effect and click Motion Paths. Choose Up, Down, Left, Right, Diagonal Down Right, or Diagonal Up Right. You can also choose to draw a custom path.

Sophistication comes with a price: complexity. When you incorporate the Motion Path option in your presentation, you will soon discover that moving an object up, down, sideways, or diagonally is relatively easy. You will also discover that once you try this powerful effect, you will want to extend its variations. That's when you will want to move beyond the basic preset Motion Paths options to Draw Custom Path with a Line, a Curve, or a Freeform. Here is where the Motion Paths effect becomes more complex. But stay the course; try, experiment, and learn. You will soon find that your efforts will produce dynamic animation that expresses your message.

Animation and the Presenter

Although the focus of this book is on the story and graphics rather than delivery skills, it is important, before concluding, to touch briefly on the presenter's narrative and body language as they relate to slide show animation. This is a skill called Graphics Synchronization and is covered in great detail in *The Power Presenter*. For now, please keep in mind that the highly sensitive optic nerves in your audience's eyes cause them to react *involuntarily* to light and motion. The instant your animation starts, *all* their attention suddenly shifts to the screen and *away* from you.

Human brains have difficulty processing multiple inputs, especially if those inputs are separate sights and sounds. Your audience is so focused on the animation, they do not listen to what you are saying, nor do they see what you are doing. Moreover, *anything* you do or say creates extra sensory data that conflicts with the enlarged activity on the presentation screen. Therefore, whenever you introduce animation on your screen, stop talking, stop moving, and allow the animation to complete its full course of action.

Whenever you introduce animation on your screen, stop talking, stop moving, and allow the animation to complete its full course of action.

"Action" is the operative word here. Action is the endgame. The action in animation is there to express your message. Your message is your call to action. The end result of all your action should be to elicit the best response from your audience. Remember Newton's Third Law of Motion: "*To every Action there is always opposed an equal Reaction*." If your animation action is disturbing, your audience will be distracted. If the dynamic action of your animation is synchronized with all the other elements of your presentation . . . your story, your graphic design, as well as your voice and body language . . . you can achieve the *ultimate* dynamic: persuading your audience to move from Point A to *your* Point B.

Chapter Thirteen

The Virtual Presentation

Company examples

- **Microsoft Office Live Meeting**
- **Cisco Systems WebEx**

Anytime, Anyplace, Anyone

A new phenomenon has entered the world of business communications: Web conferencing, or online meetings. Think of it as the *virtual presentation*. This new medium combines the Internet and related electronic technologies to enable you to present *and exchange* information and ideas with distant audiences without your physical, personal presence. In fact, you can address a group of dozens or even hundreds of audience members in different locations spread around the country or even the world. By virtue of advanced software systems developed by several companies that specialize in hosting virtual presentations, these far-flung audiences can view your slides and other graphic materials, hear your voice, and even interact with you . . . live and in real time.

The virtual presentation has many advantages. Because participants can log in from any location via the World Wide Web . . . *from the convenience of their own offices* . . . it is possible to involve an extensive universe of attendees, some of whom would otherwise be unavailable. Those who do attend online save thousands of dollars in travel costs (airfare, lodging, meals, ground transportation, and other local expenses). They also dramatically reduce their time commitment and completely eliminate the stress of travel. The presenter also realizes significant cost savings in rental fees for hotel or conference center facilities, projection and amplification equipment, and other presentation paraphernalia.

Of course, there are costs associated with the virtual presentation, depending on rates and other variables charged by the Web conferencing provider, but they will be far less than those of mounting a live presentation.

Current trends in business are helping to boost the adoption of virtual presentations. More and more companies, through expansion or partnering, are operating globally. Their diverse and dispersed entities have an ever-increasing need to share information . . . in meetings, seminars, workshops, briefings, demonstrations, or training . . . and to do it *collaboratively*. Moreover, when economic times are difficult, companies look for opportunities to trim costs. An obvious place to start is with travel expenditures for meetings and conferences. Business travel has been further reduced by heightened security concerns in the wake of the terrorist attacks of September 11, 2001 and by rising fuel costs. Finally, the virtual presentation is an attraction in its own right. Many companies like to use cutting-edge technology and new applications with the aim of fostering an image of themselves as innovative, forward-looking business players.

For all these reasons, the virtual presentation has become increasingly popular. However, learning to deliver a virtual presentation effectively takes time and effort. When the presenter and the audience are communicating over the Internet from separate locales, new factors come into play that can impact the outcome. This chapter provides a set of guidelines to help you realize the full potential of this exciting new business medium.

How the Virtual Presentation Works

Most businesspeople are familiar with teleconferencing, where multiple individuals in various locations speak together on a shared telephone line, but these exchanges lack the visual element. There is videoconferencing, provided by Polycom, Cisco Systems' TelePresence, and Hewlett Packard's Halo, but the cost of transmission and specialized equipment limits the broad use of this form of communication. The virtual presentation leverages the ubiquity of computers and the large installed base of Microsoft PowerPoint for its universal system. Think of the virtual presentation as teleconferencing on steroids, with visual aids . . . *as well as* interactivity . . . and all of it enabled by the power of the Internet, at a reasonable and economical cost.

Think of the virtual presentation as teleconferencing on steroids, with visual aids and interactivity.

A virtual presentation can combine a number of technologies:

- A standard Web browser, used to access the Internet address (URL) of the Web conferencing provider, such as Microsoft Live Meeting or Cisco's WebEx, and host the interactive presentation
- Streaming audio and video for Web-based video presentations
- Conventional telephony or Voice over Internet Protocol (VoIP) to provide the sound for audio-only presentations
- PowerPoint slides (or any similar electronic documents), shared over the Internet with all participants

Today, many virtual conferences are delivered exclusively over the Web, using Internet streaming audio and video instead of the older telephony-based audio. Increased bandwidth enables more robust conferences with live audio and video . . . in some cases, two-way audio/video for more interactive presentations. Just about any business or individual with a fast Internet connection can either attend virtual presentations as an audience member or offer them to a wide selection of target audiences. (Some technical issues related to bandwidth availability and firewall protection may affect how your company can use a virtual presentation service, but the Web conferencing provider can help you deal with your individual constraints and requirements. Suffice to say, the faster your Internet connection, the smoother the presentation will be.)

For less-formal virtual conferences and presentations, you can avoid the expense of a third-party provider such as Microsoft or Cisco. Instead, you can use instant-messaging programs such as AOL Instant Messenger or Yahoo! Messenger to conduct small text-based or videoconferences, sharing files between users as required. In addition, Web-based presentation programs, such as Google Presentations, let multiple users view the same presentation onscreen from any location in real time. These types of presentations don't have all the interactive bells and whistles of Live Meeting or WebEx, but they are considerably lower priced and easier to arrange.

If you're using a third-party conference service, you arrange all the details with the provider. You start by scheduling your virtual presentation with the provider, and then you invite your potential attendees via phone or email. The provider gives each participant who registers the URL for the online session, a conference call number (if telephony audio is used), and a password or code to enter the session. As the date for your webcast approaches, you may need to upload or email your PowerPoint slides to the Web conferencing provider for uploading to its website. You may also be able to give the presentation "live" from your desktop during the conference itself.

At the appointed time, you and your audience log in to the provider's site via the Internet and, if necessary, dial in to the call center via the telephone. You, the presenter, speak to your audience over the telephone, the VoIP line, or via streaming Internet audio or video. Simultaneously, you display your presentation on your computer, and your audience views it on their computers.

Audience members can react to your presentation through their computers, and you can see their responses on your computer screen. The presenter's screen, also called the console, has many more options than the audience's screens. The console options allow you to navigate through your graphics, respond to your audience's feedback, or solicit their opinions and reactions . . . in short, to *effectively manage* your presentation.

Your virtual audience participants can type their questions into their computers. You can respond live on the conference call, or share the question with the entire audience, or have an assistant type a direct answer to the participant who raised the question.

The virtual presentation platform offers a variety of companion features:

- **Annotation.** The presenter can highlight a particular word, number, or object on any slide using an electronic marker controlled by the mouse. This annotation feature is very much like the "telestrator" that television sports broadcasters use to draw a circle or underline on the instant replay video of a critical moment in a game. Just as the broadcaster goes on to analyze and add color to the play, you can proceed to add value to your presentation.
- **Virtual whiteboard.** The presenter can write notes or sketch diagrams that appear on the audience's computer screens, just as they would on a conference room whiteboard.
- **Application sharing.** The presenter and the audience can access software demonstrations by sharing the presenter's computer desktop. Whatever occurs on the presenter's host computer occurs on the audience's computers, just as if the software application were running on their own computers.

- **Instant polling.** The presenter can create multiple-choice questions and poll the audience about their opinions or demographics. The Web provider tabulates the results instantly, and the presenter can either share them with the audience or save them for later market analysis and research.
- **Audience attendance and reaction display.** The presenter's console has an image that provides a stylized view of the virtual audience in the virtual auditorium: a group of lighted figures arranged in rows. As the audience members key their responses into their computers, the figures change color to represent their various reactions. Red indicates "You're moving through this topic too quickly" or "I don't understand!" Green indicates "I got it!" or "*Aha!*"
- **One- and two-way videoconferencing.** Using a simple webcam, the presenter can appear via streaming audio/video on the participants' computer screens. Some providers offer multipoint video, where remote participants can also appear via webcam, for enhanced face-to-face interaction.
- **Live chat.** Audience members can submit questions via instant messaging, or chat among themselves in an ad hoc Web-based chat room.

The companies that provide virtual presentations maintain the host software required to perform these potent features. They also manage the websites, facilitate the interactions, and provide technical and administrative support to their clients.

Currently, virtual presentation technology is used primarily for three kinds of meetings:

- **Small groups** of 5 to 10 people, often for sales pitches, but also for ensemble brainstorming, problem-solving, or project management
- **Medium-sized assemblages** involving 20 to 30 people, often for training or marketing
- **Large forums** with audiences of 300 to 1,000 for major product releases, company-wide announcements, investor relations meetings, or press briefings

Quite naturally, companies in the high-tech sector have been the most enthusiastic adopters of virtual presentation technology. But many other businesses are now getting into the virtual act, including global financial firms, manufacturing companies, and service businesses.

This powerful technology is also being used for some highly specialized meetings. For example, virtual collaboration tools make it possible for a team of product designers to meet in a virtual space and work together on the same set of blueprints, even when one of the designers is in New York, a second in Los Angeles, and a third in Milan. In this chapter, we'll focus only on the one-to-many presentation, the virtual version of the mission-critical presentation *any* businessperson might have to deliver to *any* audience of *any* size gathered in cyberspace.

Preparing for Your Virtual Presentation

In most ways, preparing for a virtual presentation is just like preparing for an in-person, live presentation. All the principles you've learned throughout this book apply to the virtual presentation:

- Define your Point B and your audience's WIIFY.
- Distill your ideas into a few central themes or Roman Columns.
- Organize the Roman Columns into a logical flow.
- Illustrate your concepts with graphics that follow basic design and continuity principles you read about in Chapters 6–9.
- Practice your presentation using Verbalization.
- Deliver your presentation using Customization to make it fresh, specific, and alive.

All the principles you've learned throughout this book apply to the virtual presentation.

All these techniques are as important in the virtual presentation as they are in a live presentation. However, you must consider some important added elements when preparing for an online presentation . . . especially the first few times you approach the virtual podium:

Attend virtual presentations by others. The best way to become accustomed to the "feel" of the virtual presentation is to attend other online meetings. Log in to several virtual meetings given by others in your industry, or attend open Web meetings sponsored by Web conferencing providers. See what you like and don't like about the virtual presentations you attend: Which verbal, visual, and other techniques do the presenters use that you find effective, clear, and compelling? Which techniques fall flat? Take notes and develop your own virtual presentation style.

Listen to commercial or public radio. Listen with increased perceptivity. Since radio programming is essentially broadcasting without visual images, it serves as a metaphor for your virtual presentation, which is essentially broadcasting without visual connections. Listen to how radio professionals create visual imagery using descriptive language to paint imaginative verbal pictures. A short radio documentary on National Public Radio or even a report from the field on a news station can stimulate ideas for your own online narration.

Check your graphics. Design your virtual presentation slides using the techniques in Chapters 6, 7, and 8 and, in particular, Chapter 9, with its recommendations to use graphics to help express the continuity of your story. Without your immediate physical presence, your virtual audience will need additional guidance to navigate the flow.

After you've crafted and refined your slides, upload or send them to the Web conferencing provider following its protocols. Be sure to check how your graphics work on the screen. Make certain that your slides appear in the sequence you intend; learn how to move from one slide to the next and how to navigate around slides. If any one of your slides appears distracting when it is displayed on your screen over the Web, modify it until it works.

Verbalize for flow and timing. Remember from Chapter 10 that Verbalization crystallizes the flow of your presentation. While your graphics can help express continuity, your narration can help even further. Since you will be webcasting to an unseen audience whose immediate reactions and intake you won't be able to gauge, it is vital that you present your ideas in a clear, logical progression. Verbalization will help you know your material cold and deliver it with precision.

Verbalization also enables you to determine the running time of your presentation. In much the same way a newspaper publishes a TV schedule, the Web conferencing provider will post the duration of your virtual session. You don't want your audience exiting your virtual presentation for another program.

Get ready well in advance of showtime. On the day of your presentation, clear your schedule for at least an hour prior to the scheduled start. Review your presentation materials, your slides, your flow, and your key messages. Gather any reference or backup materials you might need, such as market research data, press commentary, or links to related websites.

As zero hour approaches, make a stop in the restroom. On your way back to your office, bring a bottle or two of drinking water to have available to moisten your mouth and throat. (Use the sport-top bottles to avoid accidental spills on your computer or notes.)

Make sure that outside noises won't interfere with your presentation. Tell the contractors doing construction in the adjacent office to take a long coffee break. Lock your office door to avoid interruptions. Then, log on to both the phone and Web conference lines about 15 minutes in advance. That way, you'll feel relaxed and comfortable when your audience begins to "arrive."

Connecting with Your Invisible Audience

In many ways, the virtual presentation is easier and more efficient than the traditional live presentation. However, there is one main disadvantage: the lack of immediate visual connection to your audience. Even if your presentation includes a live video link to the remote conference room, you can't greet your audience personally as they arrive; you can't shake their hands; you can't make eye contact or exchange smiles. While these small, human links may seem like minor niceties, they go a long way toward creating a sense of warmth and connectivity that paves the way for persuasion.

Fortunately, there are ways to compensate for the missing human elements in the virtual presentation. Here are some techniques you can use:

Use polling to get to know your audience. This is especially important when you're presenting to a large audience, and you should do it early in the presentation. Soon after you begin, ask three or four multiple-choice questions designed to sketch a quick portrait of your audience. If you are presenting a new business product or service to an array of potential customers from many different companies around the country, you might say:

> *Let's take a moment to find out a little about you. Please answer this polling question about your job description: If you're in sales or marketing, pick choice A; in human resources, pick choice B; in planning, pick choice C; and in any other department, pick choice D. [Pause while the audience makes its selections, and wait until the tally appears on your screen.] Okay, I see that most of you are in sales and marketing . . . just a few in the other areas. That will help me tailor today's presentation for your benefit. Now, please tell me about the size of your company. Choose A if your annual revenues are under $10 million, B if they range between $10 million and $50 million . . .*

These polls provide benefits both to you, the presenter, and to your audience. They help you understand your audience and adjust your presentation accordingly. They also help each member of your audience see how he or she fits in with the others in the virtual conference room or auditorium. (When you attend a meeting, you glance around the room to see who else is present, don't you? The polling feature lets your invisible audience members do the same . . . virtually.) By sharing the poll results with the dispersed attendees, you create a sense of community among all of them. Continue to poll throughout your virtual presentation. By checking the pulse of your audience periodically, you sustain the vital lifeline of persuasion . . . the "co" in communication.

By checking the pulse of your audience periodically, you sustain the vital lifeline of persuasion . . . the "co" in communication.

Invite questions and comments. Pause periodically during your presentation to give your audience a chance to absorb your ideas. This is especially useful after you've covered a particularly complex or important subject. Invite your audience to share any questions or comments they might have. Choose the best questions to answer "on the air." Based on your experience with the topics of your presentation, you should have an idea which questions recur most often and therefore represent widespread interest. Focus on these topics, and reserve the more narrowly based questions for "off the air" responses after the Web session.

Have an assistant on hand to help monitor and manage the program. Rather than trying to handle the entire online presentation on your own, have a knowledgeable assistant join you (off-camera, of course). You'll find that he or she can help you in many ways. The assistant can field questions as they come in from the audience, select which ones you respond to directly and which ones to defer, or even answer some on his or her own. The assistant can also write you a quick note to remind you of a point you may have omitted.

Creating a Winning Virtual Presentation

All of the preceding advice will help you optimize the tools and techniques necessary to make any virtual presentation. Here are some additional techniques to raise the bar and help your virtual presentation win:

Use the enrollment report to customize your virtual presentation. The Web conferencing provider will give you a log containing the name and company of each participant who registers for your meeting. This data, along with the polling and inbound questions, serve as an excellent source for the Customization techniques you learned in Chapter 11:

- **Direct References** to your participants by name.
- **Mutual References** to well-known associates of the participants or to other participants in the online session.
- **Back References** to moments or subjects that occur after your session begins.

While you're at it, incorporate the Customization techniques that require advance development and preparation:

- **Contemporize** by referring to what is happening on the day of your online presentation. Keep in mind that your virtual presentation may be attended live across multiple time zones, so replace any references to "this morning" or "this afternoon" with "today." Also keep in mind that your virtual presentation will be archived for viewing at a later date. However, since the archived version is, in effect, date-stamped, your future audiences will understand and accept the time context.
- **Localize** by referring to related aspects of any of the attendees' venues, such as well-known events, places, or businesses.
- **Data.** Provide current statistics or facts that support your message.
- **Opening Graphic.** Start your virtual presentation with a slide that includes your audience, the subject, and the date.

I repeat these techniques with the same caveat as when I introduced them in Chapter 11: Of the many presentation methods I've provided to my clients over the past 20 years, the Customization techniques are implemented the least. Yet they provide potentially the biggest bang for your buck. As important as Customization is in an in-person presentation, it becomes even more important in the virtual presentation because of the separation between you and your audience. Although you can't see all the individuals in your audience, they are just as interested in making a personal connection with you as any live audience is . . . perhaps even more so. It is your job to create the illusion of the first time, to create the sense that you and the audience are together in the same "space." All it takes is concentration. Being "live and in person" is the ultimate form of Audience Advocacy. Learn the Customization techniques and implement them.

Being "live and in person" is the ultimate form of Audience Advocacy.

Visualize your audience. With the enrollment log, polling, and inbound questions providing a snapshot of your audience, you can consciously imagine them listening to you and watching your slides click by on their screens. You can even visualize yourself in a traditional conference room or auditorium, with your audience seated before you. Imagine them absorbing your words, nodding with understanding as you explain your concepts, and smiling appreciatively at your WIIFYs. Visualizing your audience will trigger the adrenaline flow you need to present with energy and focus. It will help you avoid the feeling of isolation that might otherwise arise from being alone in your office, speaking "into the air."

Be prepared to adjust your content on-the-fly. The only constant in life is change. As so often happens in live presentations, the audience you anticipate may not be the audience that shows up. The people who register for your virtual presentation may drop out, or they may play musical chairs. For such circumstances, prepare extra material, and then use the polling comments to guide you in adjusting your content. If you have three alternative examples to illustrate a key idea, you can select the one that works best for the audience that does attend and omit the other examples. Spend more time on the topics that are important to the particular mix of audience members assembled; move swiftly through topics of lesser interest to them, or drop those topics completely.

One of the advantages of the virtual presentation is that you can navigate through and around your presentation at will. One of the windows in the presenter's console (which your audience cannot see) lists all your slides. You can move backward or forward or skip around to clarify a point in response to an audience question.

Focus on your voice. The main element missing from the virtual presentation is your personal presence. If you're giving an audio-only presentation, the audience cannot see you; they can't observe your gestures, respond to your facial expressions, or react when you look at individual members of the audience. Adjust your speaking style to make up for this missing element. Use your voice to convey absent visual impressions. Modulate your voice to emphasize your ideas. Raise and lower your pitch for variety, and when you have an especially important point to make, hit it hard. Control the rate at which you speak, pause to emphasize key ideas, and use a varied cadence to avoid the hypnotic metronome effect that many people often fall into when they speak on the telephone. (You can find a fuller discussion of vocal volume, inflection, and cadence in *The Power Presenter*.)

To optimize your vocal instrument, stand while you present. This will expand your chest and provide more air to flow through your lungs. Sitting (or worse, slouching) has the opposite effect of constricting your lungs and your air supply. Will Flash, the Broadcast Host of Microsoft Office Live Meeting, presents multiple times daily. Will has replaced his conventional office desk with an elevated one so that he can stand during his virtual presentation and access his computer at eye level.

The average businessperson speaks in a more monotonous, flat tone than he or she realizes. Heighten your vocal shadings and variations when you are "on the air." Don't become exaggerated or unnatural, but if you expand your vocal range by 10 to 20 percent, it will help convey your ideas across cyberspace. To develop facility with your vocal techniques, use a voice recorder during your Verbalization practice to hear how you sound objectively.

Use a headset microphone. For the best sound quality and the most comfortable presentation in a telephone-based conference, use a plug-in telephone headset rather than a conventional handset phone or its speakerphone feature. A conventional handset is a nuisance to hold and handle. A speakerphone distorts the sound, and, if you move your head a foot or two, you may drop out of range. A lightweight headset frees your hands to operate your computer mouse and handle your notes. It also allows you to stand and increase your lung capacity and, in that way, your vocal quality.

Even better, consider investing in a professional headset microphone or a lavaliere, also known as a lapel microphone. Like a telephone headset, a headset microphone puts the microphone directly in front of your mouth for better audio quality and frees you to move. A professional microphone, however, does this with better audio quality than a conventional consumer telephone headset. A lapel microphone provides much the same benefit, as it clips onto your lapel, the front of your blouse, your shirt, or your necktie.

Plan for video success. The typical videoconference uses a simple computer webcam, which has both good and bad aspects. The good is the price ($100 or less), the convenience, and the ease of connection and use; the bad is the overall video quality. Most webcams have relatively low resolution, resulting in inferior images compared to standard television. Therefore, dress in plain, muted colors (no distracting patterns), and position yourself against a contrasting single-color wall or backdrop. Equally important, because action captured by webcam can appear choppy on the receiving end, keep your movements to a minimum. Keep in mind, too, that your participants will see you in a small window on their computer screens, so confine your gestures to a narrow area.

When presenting, look at and speak directly to the camera. Think of the camera as a single audience member; and direct all your comments to that one person. Use the slides on your computer screen as a teleprompter so that you won't have to move your eyes too far from the lens. Try to avoid frequent glancing down at your notes, which is more distracting in video presentations than in live ones.

The Future of the Virtual Presentation

Like all new technologies, the online meeting is gaining acceptance gradually. As more people become experienced in attending and running such events, as more companies become Web-enabled, as the Internet infrastructure takes root, and as more industries replace in-person gatherings with online communication, the virtual presentation will inevitably become more pervasive. It is only a matter of *when*, not *if*.

Some may decry the advent of virtual exchanges as reducing the human element in business. Not likely. There is no substitute for personal contact when significant relationships or partnerships are at stake. There will always be a need for face-to-face meetings. People enjoy the sheer energy of socialization, with its mix of meeting and greeting, information sharing, working the room, industry gossip, and schmoozing. No matter how highly networked businesses become, the traditional sales meeting, the industry convention, the executive conference, and the in-person PowerPoint presentation will always have a place in our culture.

Nevertheless, it is clear that the virtual presentation is increasingly becoming the medium of choice for many of our everyday business communication exchanges. Master the techniques involved, and you'll become a master of this powerful new art.

Chapter Fourteen

Pitching in the Majors

Company example:

- **Microsoft**

End with the Beginning in Mind

In Chapter 10, "Bringing Your Story to Life," you read abut the Bookends linkage: start your presentation with an anecdote or data point, and *never* mention that item again until your conclusion. The two references then serve as symmetrical brackets for your presentation. Once again, I'll practice what I preach: In the Introduction, I recounted my experience with Jeff Raikes at Microsoft; and so, in the style of a Puccini opera or Broadway musical, let's reprise that anecdote.

I spent an entire session working with Jeff to focus and organize his story, and never once addressed his vocal inflection or gestures. But that focus paid off. Jeff's delivery of his presentation rang with conviction, *because getting his story right gave him the confidence to express himself with assurance*. The lesson: A well-prepared story enhances a presenter's delivery skills.

A well-prepared story enhances a presenter's delivery skills.

It All Starts with Your Story

Jeff Raikes' case in point applies to every presenter. Persuading your audience is, above all, a matter of being prepared with a logical, compelling, relevant, and well-articulated story. You need to know where you want to lead your audience, your *Point B*, and how you will get them there. To do that, you need to *focus* on the most essential elements of your story, arrange them in a lucid *flow*, and convey them with frequent emphasis on the specific benefits, the *WIIFYs*, your audience will enjoy if they heed your call to action. Finally, you need to support your presentation with *Less Is More* graphics that express your ideas clearly. Your audience is then free to concentrate on you and your discussion . . . *Presenter Focus*.

If you prepare your presentation with all these elements at their optimum, a remarkable thing happens: Your speaking skills improve significantly. Knowing that you are fully in command of your story inevitably enhances your self-confidence and poise, resulting in a far more polished and convincing presentation. That's the discipline by which most persuasive battles are won.

Are the specifics of the presenter's voice and body language important? Absolutely! Think of the diligently crafted story and graphics as a highly sophisticated communications satellite, and the presenter as a powerful Atlas rocket. NASA spends millions of dollars and thousands of hours building such satellites. If the rocket, the delivery system, is defective, the satellite doesn't go into orbit. The same is true of the presentation. If the messenger is defective, the message goes awry. The well-designed substance needs an effective delivery style to lift the payload into orbit.

However, if I were to ask you to work on your delivery style before your story and graphics were "baked," it would be like asking you to swim with your feet tied together. For this reason, I have deliberately focused on the clarity of your content in this book and have separated your delivery skills into a separate book, *The Power Presenter.*

There are several other reasons for this partitioning. In the first place, businesspeople are not performers, nor are they receptive to being treated as performers. If you focused on your delivery skills upfront, you'd feel pressured to perform. If instead you focus on your story first, you can approach those skills in a natural and conversational manner.

This raises another subtle but important point: When most normal businesspeople stand to speak, they experience a sudden rush of performance anxiety. A study cited in *The Book of Lists*, by David Wallechinsky, ranks speaking before a group as the highest anxiety-provoking event, topping fear of heights, insects, flying, and even death. The human body's instinctive reaction to anxiety is increased adrenaline flow, which makes any possible natural behavior almost impossible. Fight or flight follows. But clarity of mind diminishes performance anxiety.

Clarity of mind diminishes performance anxiety.

Another reason for excluding delivery skills from this book has to do with how the presenter relates to the graphics. Because most slides are all too often guilty of the Data Dump sin, they create a negative effect on the delivery skills: The presenter is forced to turn to the screen and read the slides verbatim, thus evoking resentment from the audience. They think, "I can read it myself!"

The solution is to coordinate the delivery skills, the slide design, and the narrative. This is a unique skill, called Graphics Synchronization that you can find in the pages of *The Power Presenter*, but even there, the essential starting point is effective graphic design. If you design your slides using the Less Is More principle, you will be free to concentrate on addressing the audience directly and adding narrative value . . . further support for Presenter Focus.

Practice, Practice, Practice

Your sense of self-confidence and your ability to persuade will be even further enhanced if you devote sufficient time to polishing and practicing your presentation after you've done your basic preparation. This means Verbalization, one of the essential keys to making your presentation truly effective. Simply put, the more time you allot to Verbalization, and to its vital counterpart, Spaced Learning (the opposite of cramming), the better your presentation will be.

Spaced Learning will help you take command of your presentation and make you more confident when you deliver it. You will be less rushed and more poised. Spaced Learning will also help you give your presentation greater polish. Each time you practice Verbalization, you will find new ways to clarify your story. Finally, Spaced Learning will give you the time to make your presentation more succinct: to trim deadwood, eliminate interesting but irrelevant detail, and hone your fine points. The result will be a finely sharpened message.

Here's a checklist for your practice:

- Verbalize your presentation repeatedly. You can do this alone or into an audio . . . *not* video . . . tape recorder. Video recorders make you self-conscious about your appearance; audio-only recordings allow you to concentrate on your narrative. You can also Verbalize in front of a trial audience of colleagues or friends. In all cases, Verbalize *with* your slides.
- Time your presentation to be certain that it works effectively within the allotted time period.
- Each time you Verbalize, use Internal Linkages to connect your ideas and External Linkages to connect with your audience.
- Pay careful attention to your Verbiage. Speak positively and always as an Audience Advocate.

Make good presentation practices habitual so that you employ them almost without thinking. Verbalization and Spaced Learning are enormously powerful tools that enhance the effectiveness of every other technique you've learned in these pages. Practice them!

Every Audience, Every Time

You've seen many of the concepts in this book illustrated with examples from my work with companies preparing for their IPO road shows. Presenting an IPO road show is the ultimate in mission-critical assignments: It's like piloting a space shuttle flight, conducting the New York Philharmonic, competing in the Olympic finals, or pitching the seventh game of the World Series. I hope you've found these IPO stories to be clear illustrations of the Power Presentations techniques.

More importantly, I hope you've recognized that these very same techniques can and should be used in *every* presentation or speech. Their fundamentals have existed since Aristotle. They have been used by Abraham Lincoln, Winston Churchill, John F. Kennedy, and Martin Luther King, Jr. You can use them too.

You may have a long and successful business career and never have the opportunity to present in an IPO road show. You may work in a field that isn't usually considered part of the business world: in a government agency, a community organization, or a not-for-profit group that provides many important and valuable services to our society. You may even work on a strictly unpaid, volunteer basis, perhaps as an officer of the Rotary club, the school board, or a charitable foundation.

No matter what kind of presentations you make, no matter where you make them, and no matter who your audiences are, you want to make your presentations as powerful and as persuasive as possible. The challenge is as urgent as convincing your company's executive team to support your new business idea, as close to home as winning your neighbor's support in a local election, or as personal as capturing the delighted interest of a classroom of first-graders on Open School day. In every case, your results are on the line.

If a presentation is worth doing at all, it's worth doing well. Be all that you can be. Invest the time and energy to make *every* presentation a Power Presentation.

You may never get to pitch in the seventh game of the World Series, but *every* time you set out to persuade, you're pitching in the majors. Go for the win!

Appendix A

Tools of the Trade

The Presentation Environment

In the theater, the finest play by the finest playwright, from William Shakespeare to Arthur Miller, can be affected . . . for good or for bad . . . by the staging. So it is with a presentation. You could create an effective Power Presentation with a lucid story and vivid graphics that uses every technique you learned in this book, only to have it all diminished by the presentation environment.

As the presenter, you bear the ultimate responsibility for your presentation. To ensure that you and your graphics have the maximum impact on your audience, it is your job to optimize the environment. Here's a checklist:

- **Familiarization.** Arrive early and walk the entire presentation environment, not just the stage. Go to each part of the room and check the sight lines. Check, double-check, and triple-check everything.
- **Equipment.** Have a backup for each piece of technical equipment: computer, video, product demonstration, and projector. Remember Murphy's Law: "Anything that can go wrong, will." Also remember its corollary, Sullivan's Law: "Murphy was an optimist."
- **Amplification.** Check the sound system and test the microphone. Most presenters need amplification with an audience of 50 or more. If you are soft-spoken, use a microphone with an audience of 25 or more.
- **Projection screen.** Present with the screen to your left as you face the audience. In the discussion of Perception Psychology in Chapter 6, "Communicating Visually," you learned that audiences in Western culture find it natural to move from left to right. With this arrangement, every time you click to a new slide, the audience travels from you to the screen and across the image easily and comfortably.

- **Lighting.** Keep the illumination low enough to create contrast on the screen, but never so dark that you lose eye contact with the members of your audience.
- **Pointers.** Lasers, retractable metal rods, lighted arrows, and saber-like wooden lances . . . all weapons . . . must be checked at the door. They are more hindrances than aids.
- **Timing.** Ask someone you know to sit in the audience during your presentation and send you countdown signals so that you finish in your allotted time.
- **Liquids.** Drink water to moisten your mouth. Avoid milk and milk products as they coat the throat with a film. Avoid carbonated beverages.
- **Attire.** "When in Rome, do as the Romans do." Wear clothing appropriate to the event: suits for business occasions, and casual dress for informal occasions. Men: Button your suit jackets. Women: Leave your clanging or glittering jewelry at home.

Unfortunately, the world is not perfect. The best hotel banquet salons, the most modern executive briefing centers, and even the slickest conference rooms equipped with the newest, most expensive, highest-end equipment need to be fine-tuned. In the presentation world, the optimal is sometimes unattainable. If so, live with it.

If you overcome adversity, your audience will empathize with you, appreciate your effort, and value you more.

If the good Lord gives you a lemon, make lemonade!

Presentation Checklists

The Four Critical Questions

1. What is your **Point B**?
2. Who is your audience, and what is their **WIIFY**?
3. What are your **Roman Columns**?
4. Why have you put the Roman Columns in a particular order? In other words, which **Flow Structure** have you chosen?

WIIFY Triggers

1. "This is important to you because . . . " (The presenter fills in the blank.)
2. "What does this mean to you?" (The presenter explains.)
3. "Why am I telling you this?" (The presenter explains.)
4. "Who cares?" ("You should care, because . . . ")
5. "So what?" ("Here's what . . . ")
6. "And . . . ?" ("Here's the WIIFY . . . ")

Seven Classic Opening Gambits

1. **Question.** A question directed at the members of the audience.
2. **Factoid.** A striking statistic or little-known fact.
3. **Retrospective/prospective.** A look backward or forward.
4. **Anecdote.** A short human-interest story.

5. **Quotation.** An endorsement about your business from a respected source.
6. **Aphorism.** A familiar saying.
7. **Analogy.** A comparison between two seemingly unrelated items that helps illuminate a complex, arcane, or obscure topic.

Sixteen Flow Structures

1. **Modular.** A sequence of similar parts, units, or components in which the order of the units is interchangeable.
2. **Chronological.** Organizes clusters of ideas along a timeline, reflecting events in the order in which they occurred or might occur.
3. **Physical.** Organizes clusters of ideas according to their physical or geographic location.
4. **Spatial.** Organizes ideas conceptually, according to a physical metaphor or analogy, providing a spatial arrangement of your topics.
5. **Problem/solution.** Organizes the presentation around a problem and the solution offered by you or your company.
6. **Issues/actions.** Organizes the presentation around one or more issues and the actions you propose to address them.
7. **Opportunity/leverage.** Organizes the presentation around a business opportunity and the leverage you or your company will implement to take advantage of it.
8. **Form/function.** Organizes the presentation around a single central business concept, method, or technology, with multiple applications or functions emanating from that central core.
9. **Features/benefits.** Organizes the presentation around a series of your product or service features and the concrete benefits provided by those features.
10. **Case study.** A narrative recounting of how you or your company solved a particular problem or met the needs of a particular client and, in the telling, covers all the aspects of your business and its environment.
11. **Argument/fallacy.** Raises arguments against your own case, and then rebuts them by pointing out the fallacies (or false beliefs) that underlie them.

12. **Compare/contrast.** Organizes the presentation around a series of comparisons that illustrate the differences between your company and other companies.
13. **Matrix.** Uses a two-by-two or larger diagram to organize a complex set of concepts into an easy-to-digest, easy-to-follow, and easy-to-remember form.
14. **Parallel tracks.** Drills down into a series of related ideas, with an identical set of subsets for each idea.
15. **Rhetorical questions.** Asks, and then answers, questions that are likely to be foremost in the minds of your audience.
16. **Numeric.** Enumerates a series of loosely connected ideas, facts, or arguments.

Twelve Internal Linkages

1. **Reference the Flow Structure.** Make repeated references to your primary Flow Structure as you track through your presentation.
2. **Logical transition.** Close your outbound subject; lead in to your inbound subject.
3. **Cross-reference.** Make forward and backward references to other subjects in your presentation.
4. **Rhetorical question.** Pose a relevant question, and then provide the answer.
5. **Recurring theme.** Establish an example or data point early in your presentation, and then make several references to it throughout your presentation.
6. **Bookends.** Establish an example or data point early in your presentation, and never mention it again until the end.
7. **Mantra.** Use a catchphrase or slogan repeatedly.
8. **Internal summary.** Pause at major transitions and recapitulate.
9. **Enumeration.** Present related concepts as a group, and count down through each of them.
10. **Do the math.** Put numeric information in perspective.

11. **Reinforce Point B.** Restate your call to action at several points throughout your presentation.
12. **Say your company name.** State your company, product, or service name often.

Seven External Linkages

1. **Direct reference.** Mention specifically, by name, one or more members of your audience.
2. **Mutual reference.** Make reference to a person, company, or organization related to both you and your audience.
3. **Ask questions.** Address a question directly to one or more members of your audience.
4. **Contemporize.** Make reference to what is happening today.
5. **Localize.** Make reference to the venue of your presentation.
6. **Data.** Make reference to current information that links to and supports your message.
7. **Customized opening graphic.** Start your presentation with a slide that includes your audience, the location, and the date.

Five Graphic Continuity Techniques

1. **Bumper slides.** Graphic dividers inserted between major sections of the presentation to serve as clean, quick, and simple transitions.
2. **Indexing/color coding.** Uses a recurring object as an index, highlighted in different colors to map the different sections of a longer presentation.
3. **Icons.** Express relationships among ideas using recognizable symbolic representations.
4. **Anchor objects.** Create continuity with a recurring image that is an integral part of the illustration.
5. **Anticipation space.** Uses empty areas that are subsequently filled, setting up and then fulfilling subliminal expectations.

Acknowledgments

First and foremost, as on the first page of the Introduction, and on every other page, and in a straight line stretching all the way back to the Stanford-versus-UCLA football game in 1956 (Stanford 13, UCLA 14, unfortunately), where I first met him, is Benji (as those of us who love him call him) Rosen. What I said on the first page is well worth repeating: The conversation I had with Benji 31 years (almost to the day) after that football game changed my life. For that, I am eternally grateful.

I wouldn't have met Benji, however, nor would I have met all the other wonderful people in this book (in fact, I might have been a dentist in Great Neck, New York), if I hadn't walked into the freshman speech class at New York University, taught by one of Professor Ormond Drake's brightest young Turks, Harry Miles Muheim. From the moment I met him, Harry's eloquence, intellect, charm, and wit made me a starstruck acolyte. Whatever Harry said, did, or referenced became my gospel. Within weeks of the first class, my intended dental career receded into the dusty rafters of the little theatre group that Harry suggested I join. That experience became the seed for a career that led all the way to the CBS Broadcast Center in Manhattan, with a major stop at the Stanford University Graduate School of Speech and Drama, of which Harry was an alumnus. But Harry was far more than a mentor; he remained my role model and friend throughout the many stages of my career. Thank you, Mu.

Special thanks to Tim Moore, my publisher at Financial Times/Prentice Hall, and to Karl Weber (and to Linda Chester, who introduced me to Karl), who helped me convert the spoken words and PowerPoint slides of my Power Presentations program into text. Tim and Karl brought a combination of creative synergy and professional acumen I have not experienced in all my years and all my careers. Karl contributed his considerable business and publishing expertise while sharing the insights and resonance of a fellow ink-stained wretch. Tim contributed his wisdom and most welcome wit, as well as a stellar team of publishing professionals, including Russ Hall, Gail Cocker-Bogusz, and especially Donna Cullen-Dolce. For the revised version, Gayle Johnson, Gloria Schurick, and Lori Lyons continued the team's professionalism.

My connection to Tim extends back along the lines that weave through Cisco Systems, beginning with Don Valentine at the IPO, and continuing through hundreds of Cisco people, culminating with Sue Bostrom and Jim LeValley. Most prominent in that lengthy chain are Chuck Elliot, Corinne Marsolier, Philippe

Brawerman, and Brent Bilger. The company that internetworks all networks knows how to grab, navigate, and deposit.

I am grateful to all the fine young women who have helped me run the business side of Power Presentations and, in so doing, freed me to develop the creative concepts that led to this book: Jennifer Haydon, Nancy Price, Susan Hill, Heather Scott, Jennifer Turcotte, and Nichole Nears. Jennifer, the penultimate, has been the ultimate researcher, finding facts and people with the diligence of a Scotland Yard detective. Nichole, the latest in the illustrious line, not only managed the many drafts of the electronic manuscript, she also single-handedly regenerated every graphic element that took me a decade and a half to create in PowerPoint.

Thanks too to Jim Welch, the *n*th power behind the Power; to Bill Davidow, a man with the uncanny dexterity to see the forest and the trees simultaneously; to Don Valentine, whose own ability to tell a story and facility with Less Is More makes him as quotable as Hamlet; to Melvin Van Peebles, the only man I know who can brainstorm in midair; to Christopher Spray, who helped me seek out my own Point B; to Quentin Hardy, who provided the seed; to B. J. Coffman, for the spark; to Cousin Joel Goldberg, who put the vision in television; to Warren Kaplan, in his nature; to Shelley Floyd, always there; to Roberta Baron and Denise Burrows, both P-I-T-Apostrophe; and to Frank Perlroth, for the *joie de vivre*.

In *The King and I*, Anna, a teacher, sings of being taught by her students. I have been taught by my students, the thousands of businessmen and women whose daily, real-life, mission-critical communication challenges enabled me to take the esoteric concepts and practices of professional media and translate them into everyday terms. In particular, my gratitude goes to the people who granted me permission to discuss my work with them in this book: Andrea Cunningham, Don Valentine, Cate Muther, Jeff Raikes, Dan Warmenhoven, Alex Naqvi, Jim Bixby, Reed Hastings, Judy Tarabini McNulty, Chuck Geschke, Randy Steck, Dr. Robert Colwell, Dadi Perlmutter, Dave Castaldi, Dr. Nancy Chang, Jerry Rogers, Hugh Martin, Dr. Emile Loria, Scott Cook, Mike Pope, Adrian Slywotzky, James Richardson, Tim Koogle, Gary Valenzeula, Bud Colligan, Carol Case, Dr. Charles Ebert, Vince Mendillo, Jim Flautt, Jon Bromberg, Gary Stewart, Dr. Jacques Essinger, Chuck Boesenberg, Roger McNamee, Don Listwin, Will Flash, and Richard Bretschneider.

And to Lucie, the Anchor Object in my life, *Sabor a Mi*.

About the Author

Jerry Weissman, the world's #1 corporate presentations coach, founded and leads Power Presentations, Ltd. in Foster City, California. His private clients include executives at hundreds of the world's top companies, including Yahoo!, Intel, Cisco Systems, Intuit, Dolby Laboratories, and Microsoft.

Weissman coached Cisco executives before their immensely successful IPO road show. Afterward, the firm's chairman attributed at least two to three dollars of Cisco's offering price to Weissman's work. Since then, he has prepared executives for more than 500 IPO road shows, helping them raise hundreds of billions of dollars. His techniques have helped another 500 firms develop and deliver their mission-critical business presentations.

Weissman is the author of the global bestseller *Presenting to Win: The Art of Telling Your Story* (Financial Times Prentice Hall, 2003); *In the Line of Fire* (Pearson Prentice Hall, 2005) and its companion DVD, *In the Line of Fire: An Interactive Guide to Handling Tough Questions* (www.powerltd.com); and *The Power Presenter: Technique, Style, and Strategy from America's Top Speaking Coach*.

Index

A

B

E

F

G–H

M

N

O

P

Q–R

S

T

U–V

W–Z

FT Press
FINANCIAL TIMES